MATHEMATICS —
CONNECTION AND BEYOND

Yearbook 2020
Association of Mathematics Educators

Related Titles

Big Ideas in Mathematics, Yearbook 2019
edited by Tin Lam Toh and Joseph B W Yeo
ISBN: 978-981-120-536-1

Mathematics Instruction: Goals, Tasks and Activities, Yearbook 2018
edited by Pee Choon Toh and Boon Liang Chua
ISBN: 978-981-3271-66-1

Empowering Mathematics Learners, Yearbook 2017
edited by Berinderjeet Kaur and Ngan Hoe Lee
ISBN: 978-981-3224-21-6

Developing 21st Century Competencies in the Mathematics Classroom, Yearbook 2016
edited by Pee Choon Toh and Berinderjeet Kaur
ISBN: 978-981-3143-60-9

Effective Mathematics Lessons through an Eclectic Singapore Approach, Yearbook 2015
by Khoon Yoong Wong
ISBN: 978-981-4696-41-8

Learning Experiences to Promote Mathematics Learning, Yearbook 2014
edited by Pee Choon Toh, Tin Lam Toh and Berinderjeet Kaur
ISBN: 978-981-4612-90-6

Nurturing Reflective Learners in Mathematics, Yearbook 2013
edited by Berinderjeet Kaur
ISBN: 978-981-4472-74-6

Reasoning, Communication and Connections in Mathematics, Yearbook 2012
edited by Berinderjeet Kaur and Tin Lam Toh
ISBN: 978-981-4405-41-6

Assessment in the Mathematics Classroom, Yearbook 2011
edited by Berinderjeet Kaur and Khoon Yoong Wong
ISBN: 978-981-4360-97-5

Mathematical Applications and Modelling, Yearbook 2010
edited by Berinderjeet Kaur and Jaguthsing Dindyal
ISBN: 978-981-4313-33-9

Mathematical Problem Solving, Yearbook 2009
edited by Berinderjeet Kaur, Ban Har Yeap and Manu Kapur
ISBN: 978-981-4277-20-4

MATHEMATICS — CONNECTION AND BEYOND

Yearbook 2020
Association of Mathematics Educators

editors

Tin Lam Toh
Ban Heng Choy
Nanyang Technological University, Singapore

Published by

World Scientific Publishing Co. Pte. Ltd.

5 Toh Tuck Link, Singapore 596224

USA office: 27 Warren Street, Suite 401-402, Hackensack, NJ 07601

UK office: 57 Shelton Street, Covent Garden, London WC2H 9HE

Library of Congress Control Number: 2021015353

British Library Cataloguing-in-Publication Data
A catalogue record for this book is available from the British Library.

MATHEMATICS — CONNECTION AND BEYOND
Yearbook 2020 Association of Mathematics Educators

ISBN 978-981-123-697-6 (hardcover)
ISBN 978-981-123-820-8 (paperback)
ISBN 978-981-123-698-3 (ebook for institutions)
ISBN 978-981-123-699-0 (ebook for individuals)

For any available supplementary material, please visit
https://www.worldscientific.com/worldscibooks/10.1142/12279#t=suppl

Printed in Singapore

Content

Chapter 1

Mathematics – Connections and Beyond

Ban Heng CHOY, Tin Lam TOH

Mathematics is an inter-connected discipline which demands teachers to emphasize and present the subject as a connected and coherent enterprise. But how do teachers emphasize these mathematical connections and what do teachers need to know to do this ambitious work? Following the metaphor inspired by Felix Klein's seminal work *Elementary Mathematics from a Higher Standpoint*, we argue that it is critical for teachers to see mathematics, not just from a higher standpoint, but from different vantage points to emphasize these connections. In this chapter, we highlight how the various authors in this book present diverse ideas of seeing mathematics from different vantage points. In addition, we position horizon content knowledge as an essential body of knowledge beyond mathematics that teachers ought to know to empower them in shifting towards a more student-centric teaching approach in mathematics classrooms.

1 Introduction

The theme of this year book *Mathematics – Connections and Beyond* is a natural progression from the previous year book *Big Ideas in Mathematics*. Big Ideas in Mathematics is the most recent emphasis in the mathematics curriculum by the Singapore Ministry of Education. As illustrated by Hurst (2019), an emphasis on big ideas in mathematics naturally points towards appreciating connection in mathematics. Connection can be interpreted as connection across topics in mathematics; connections between mathematics and other topics; connections between mathematics and the real world, etc.

This book is a culmination of the many ideas shared during the lectures and workshops of the annual Mathematics Teacher Conference 2019, organized by the Association of Mathematics Educators. The chapters of this yearbook discuss the various interpretations of connection and contain rich discourse beyond connection.

2 Connection across various topics in mathematics

Mathematics is an integrated and inter-connected discipline and not a collection of topics, which is often presented in silos by many teachers. As highlighted by Choy (2019), there could be several reasons for this. Firstly, teachers may find it challenging to structure their instruction around key mathematical ideas because the most curricula are not explicitly organized around these ideas. Secondly, teachers will need to make sense of these ideas for themselves before they can support their students in making these connections. Thirdly, teachers are required to adopt a more dialogic stance by focusing on ways of orchestrating mathematically productive discussions about tasks. Despite these challenges, it is critical for teachers to present "mathematics as a coherent and connected enterprise" (Charles, 2005, p. 17) so that students' understanding of mathematics is enhanced. In particular, when teachers begin to emphasize these connections during teaching, it is more likely for students to "see mathematical connections in the rich interplay among mathematical topics, in contexts that relate mathematics to other subjects and in their own interests and experience" (National Council of Teachers of Mathematics, 2000, p. 4).

Doing so would require teachers to have a deep understanding of mathematics and to be able to make use of the connections between the mathematics they learned at university and school mathematics in their teaching. This, obviously, requires a different kind of knowledge from teachers, beyond the standard university mathematics. Felix Klein (1849 – 1925), a famous German mathematician and mathematics educator, wrote in a seminal series of books titled *Elementary mathematics from a higher standpoint* that the "real goal" of a mathematics teacher is "to draw (in ample measure) from the great body of knowledge taught to you here as vivid stimuli" for his or her teaching (Klein, 2016, p. 2). Klein wanted mathematics teachers to see the connections between the main problems in the various sub-fields of mathematics in relation to what they teach in school mathematics. In other words, Klein's idea was for teachers to make connections by seeing school mathematics from a higher standpoint.

Drawing on the metaphor of learning mathematics as hiking up a mountain, we argue that teachers not only need to be able to see school mathematics from a higher standpoint, but more importantly, they need to be able to see school mathematics from *different* vantage points. It is true that a higher standpoint often offers a breathtaking view for the hiker. But it is also true that one can grasp a different view of the same mountain along different lookout points along the mountain trail, and at times, views of a different mountain. Each of these views offers the hiker a more complete and enjoyable experience of the journey. Likewise, in the learning and teaching of mathematics, it is crucial for teachers to see connections between different ideas taught in school mathematics— across the same levels and beyond different levels. These connections can be highlighted by teachers at different opportune moments in the curriculum, offering students a different view of mathematics at different vantage points! Moreover, at certain vantage points, teachers can point out "other mountains that comes into their views" to students, highlighting the connections between mathematics and other subjects such as Science.

Together with this introductory chapter, there are 11 chapters in this book. The chapters are organized according to three sections: (A) discussion of connections and beyond based on empirical studies (chapters 2 to 4); (B) teaching ideas of how connections can be introduced into the mathematics classroom (chapters 5 to 9); and (C) knowledge beyond connection of mathematics (chapters 10 and 11).

3 Section A: Connections and beyond based on empirical studies

This section contains three chapters presented by researchers based on their empirical studies. Chapter 2 begins with Lee and Shin, who believe that for mathematics learning to be meaningful, numerical calculations must be connected to quantitative operations. In this chapter, Lee and Shin presented a case study of the problem-solving activity of one student, demonstrating how the ability to operate with two levels of units affect his problem-solving activities in two multiplicative problem contexts.

In Chapter 3, Toh, Cheng, Lim and Lim argue that using comics for teaching lower secondary statistics has the ability to not only motivate students to learn the subject, but also to introduce to the students the various statistical reasoning and to engage them in higher order thinking skills and stretch their critical thinking ability. In the snapshot of the classroom episode of instruction using comics from one participating teacher, it was seen that the teacher consciously connect the statistical content to the real world to establish the relevance for the students.

In Chapter 4, Kaur and Tong provide readers with a clearer picture of what happens in the mathematics classroom in Singapore. Contrary to the usual belief that mathematics teaching in Singapore schools is all about drill and practice, data from a large-scale local research, it was found that teachers go beyond the traditional drill and practice approach. Singapore teachers tend to draw from a combination of several models from both East and West for their instructional approach.

Their model of instruction, which Kaur and Tong name it "the hybrid model", make it possible for teachers to be receptive of various innovative approach beyond the traditional ones. This leads to Section B of the book.

4 Section B: Teaching ideas introducing connection

This section contains five chapters by authors who introduce practical ways to introduce mathematical connections in the mathematics classrooms. The section begins with Chapter 5 by Kissane, who cautioned the readers not to limit their attention to the utility value of mathematics, as equally important is to get the students to be attracted to the subject, that is, the aesthetic quality of the subject and its rich collection to the cultural heritage. From this perspective, Kissane introduces the connection of mathematics to "the wider world beyond its practical application."

In Chapter 6, Low and Wong proposed the use of *connecting* tasks. These tasks are meant to be used within and across levels to make connections between various representations of the same mathematical concept, connections between different mathematical concepts and procedures, and connections between different strands of mathematics.

Besides seeing mathematical concepts by making connections between procedures and representations, Yeo K. K. J. in Chapter 7 offers teachers the idea of teaching mathematics through problem solving as a way to make connections between different solutions and mathematics. Teaching through problem solving requires teachers to shift towards a more process-based approach where getting a correct answer to a problem is not the main point. Rather, it is about engaging pupils with open-ended problems by guiding them to uncover the mathematical connections embedded in the problems.

In Chapter 8, Vuong describes how the 5E model can be adapted for the mathematics classrooms to provide students opportunities to learn mathematics through making connections. The 5E model is another possible model of instruction, which focuses on guiding learners to make connections from prior knowledge. Originally intended for science lessons, the model comprises a recursive cycle of five cognitive stages in inquiry-based learning: engagement, exploration, explanation, elaboration, and evaluation.

In Chapter 9, Yap describes an approach of teaching probability first through engaging the students in making conjectures and preparing for their interpretation without first beginning by introducing technical terms. Probability simulation is introduced to facilitate students' development of interpretation of probabilistic event. This is another illustration of connecting probability to probabilistic events in the real world.

5 Section C: Knowledge beyond connection

As we have highlighted in the introduction, it is important for mathematics teachers to see the connections between the main problems in the various sub-fields of mathematics in relation to what they teach in school mathematics (Klein, 2016) or to see school mathematics from a higher standpoint.

Illustrating this idea, Yeo B. W. J., in Chapter 10, unpacked the meaning of *equivalence*, one of the eight big ideas of school mathematics recently introduced by the Singapore Ministry of Education. The author explicates through detailed exemplification that solving equations is actually about converting an equation to another equivalent one. Yeo further highlighted how equivalence can be introduced to school students without the idea of an equivalent relation.

In Chapter 11, the last chapter of this book, Ho et al. presents computational thinking as a useful way of thinking in mathematics. Four design principles were presented to design lessons that promote computational thinking to forge mathematical ideas and enhance mathematical learning.

6 Concluding Remarks: Teachers' Knowledge for Teaching

The chapters in this book highlights what it means by making connections when teaching mathematics. But what do teachers need to know to do this? Building on the notion of Pedagogical Content Knowledge (Shulman, 1986), Ball, Thames, and Phelps (2008) described three subsets of PCK: Knowledge of Content and Student (KCS); Knowledge of Content and Teaching (KCT); and Knowledge of Content and Curriculum. They also expanded the idea of subject-matter knowledge (SMK), by describing a new construct, *specialised content knowledge* (SCK), which consists of mathematical knowledge and skills that are part of the subject-matter knowledge unique to teaching. SCK distinguishes itself from *common content knowledge* (CCK) in that the former is only needed in teaching (Hill et al., 2008), and this may include knowledge of less common algorithms, different representations of concepts, and explanations of rules and procedures (Ball et al., 2008). Here, it is useful to think of common content knowledge as the corpus of knowledge often presented in a typical university course and specialised content knowledge as the body of knowledge often presented in a typical mathematics education course. However, neither CCK nor SCK would be sufficient for teachers to enact this paradigm shift of teaching approaches as highlighted by Kaur in her chapter.

The third component of SMK is horizon content knowledge, which refers to the "is an awareness of how mathematical topics are related over the span of mathematics included in the curriculum", and this may include "vision useful in seeing connections to much later mathematical ideas" (Ball et al., 2008, p. 403). This characterisation of mathematical knowledge for teaching (MKT) is represented in Figure 1.

Mathematical Knowledge for Teaching (MKT)

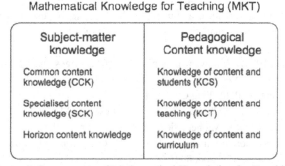

Figure 1. Components of Mathematical Knowledge for Teaching

It is this notion of horizon content knowledge which we conjecture to be critical for teachers to make these connections. For teachers to see school mathematics from different vantage points, they need to have the vision and awareness of how different topics of school mathematics are related to each other, and how these topics are related to university mathematics. However, the notion of horizon content knowledge has not been clearly defined and it remains to be seen how this body of knowledge can be characterized. For example, does this knowledge include the mathematical connections that can be applied to other disciplines beyond mathematics? Does it include an awareness of how mathematics can be applied in real-world problems? Clarifying and thinking about ways to develop teachers' understanding of this knowledge will be important as we grapple with the notion of teaching towards big ideas.

References

Ball, D. L., Thames, M. H., & Phelps, G. (2008). Content knowledge for teaching: What makes it special? *Journal of Teacher Education, 59*(5), 389-407. doi:10.1177/0022487108324554

Charles, R. I. (2005). Big ideas and understandings as the foundation for elementary and middle school mathematics. *Journal of Mathematics Education Leadership, 7*(3), 9 - 24.

Choy, B. H. (2019). Teaching towards big ideas: Challenges and opportunities. In T. L. Toh & B. W. J. Yeo (Eds.), *Big ideas in Mathematics* (pp. 95 - 112). Singapore: World Scientific Publishing Co. Pte. Ltd.

Hill, H. C., Blunk, M. L., Charalambous, C. Y., Lewis, J. M., Phelps, G. C., Sleep, L., & Ball, D. L. (2008). Mathematical knowledge for teaching and the mathematical quality of instruction: An exploratory study. *Cognition and Instruction, 26*(4), 430-511. doi:10.1080/07370000802177235

Klein, F. (2016). *Elementary mathematics from a higher standpoint: arithmetic, algebra, analysis* (G. Schubring, Trans. Vol. 1). Berlin: Springer.

National Council of Teachers of Mathematics. (2000). *Principles and standards for school mathematics*. Reston, VA: National Council of Teachers of Mathematics.

Shulman, L. S. (1986). Those who understand: Knowledge growth in teaching. *Educational Researcher, 15*(2), 4-14.

Chapter 2

Reasoning with Quantitative Units in Problem Solving: A Case of JuHa

Soo Jin LEE, Jaehong SHIN

For centuries, scholars in the field of mathematics education have researched psychological and mathematical foundations for students' development of mathematics across multiple domains. Students' ability to reason with quantitative units, especially, their ability to produce and coordinate multiple levels of units has been considered as a cognitive core, connected to their development of several mathematical concepts including counting, whole number operations, fraction, ratio and proportion, algebra, and function. In this chapter, we present a fine-grained analysis of problem-solving activities of one student, JuHa, who reasoned with two levels of units. We show how an ability to use two levels of units affords and constrains his problem-solving activities in two multiplicative problem contexts: equal sharing problem and reversible multiplicative reasoning problem.

1 Introduction

As mathematics educators, how do we prepare our students for the future society where machines do all the calculations and could pull out tons of information quite easily? What is the implication of the changing society for mathematics educators? The *Fourth Industrial Revolution* was a hot button issue worldwide when the term was announced by Klaus Schwab in the 2016 World Economic Forum Meeting in Davos, Switzerland.

Thriving in the fourth revolution requires a shift away from developing competencies that *compete with* what computers can do towards nurturing competencies that *complement* the capabilities of computers (Gravemeijer et al., 2017). Conrad Wolfram, in his TED Talk in 2010 on 'what mathematics do we teach when computers do all mathematics?', argued the following:

> We have two kinds of mathematics in the world. The first is in the real world that is a problem-solving subject that is as important as it's never been in world's history to every aspect of our lives. Then there's this mathematics in school that's completely or increasingly disconnected from that. It's that difference in the subject matter that I think is the focus we absolutely need to be honest when we talk about this. What we're mostly talking about in my view is 80% the wrong subject.

He further stressed that emphasis on numerical calculations without engaging students in meaningful problem-solving activities is one reason our kids are losing interest in mathematics.

Modelling real-life problems requires students to develop sophisticated quantitative reasoning. In other words, quantitative reasoning leads to students' involvement with more authentic mathematical modeling of real-world problems (Thompson, 2011). Moreover, students should begin to develop their quantitative reasoning early on in their mathematics education as it takes more than a year to develop (Smith & Thompson, 2007). Students' ability to reason with quantitative units, especially their ability to produce and coordinate multiple levels of units, has been considered as a cognitive core which is connected to their development of several mathematical concepts including counting, whole number multiplication and division, integer addition, fraction, ratio and proportion, algebra, and function. At the same time, numerous studies have accounted for students' inability to coordinate two and three levels of units as a root cause for poor performance across several domains in middle school mathematics including fraction arithmetic, algebraic reasoning, and integer addition (e.g., Norton, Boyce, Ulrich, & Phillips, 2015).

This chapter examines the problem-solving activities of JuHa, who reasons with two levels of units, in two different mathematical domains: equal-sharing and reversible multiplicative reasoning problem. Specifically, we illustrate how his reasoning with two levels of quantitative units afforded and constrained his problem-solving activities in the two different mathematical contexts.

2. Theoretical Constructs

2.1 *Reasoning with Quantitative Units*

Quantities are measurable properties of a person's concept of an object or a phenomenon, and a measurable property involves a measurement unit and a measurement process (Thompson, 2011). To illustrate, height is a measurable property of a person. We can talk about it without knowing how tall that person is. Even when we do not know a value for a quantity, we can still imagine a measurement unit and measurement process that could be used to find a value for a person's height, such as subdividing the height into centimeters (a measurement unit) and counting them up (a measurement process). Likewise, some students might view the bar in Figure 1 as an object with length measurement. Depending on a measurement unit and process taken with the measurement unit, they could quantify the bar in various ways.

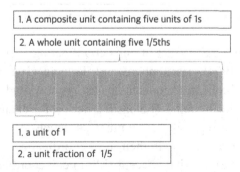

Figure 1. Two different ways of reasoning with quantitative units with bar models.

2.2 *Units Coordination and Three Multiplicative Concepts*

One of the key constructs for students' reasoning with quantitative units that researchers have been attending to is the students' construction of multiple levels of units and coordination of them in particular problem situations (Hackenberg & Tillema, 2009; Norton & Boyce, 2015; Steffe, 1994). A *units coordination* involves two composite units. A student engaged in the units coordinating activity distributes the units of one composite unit across the elements of another composite unit (Steffe, 1992; 1994). Assimilating a situation with two levels of units means a student can simultaneously recognize two levels of units and organize them with a units coordinating structure (Norton & Boyce, 2015). When a student assimilates a situation with three levels of units, he or she can attend to three levels of units concurrently, which makes it possible for the student to transform a three-levels-of-units structure to the other. To illustrate, if students can conceive of 35 by coordinating three levels of units, they can think of 35 as a unit composed of five units, each composed of seven units of 1 at the outset (see Figure 2).

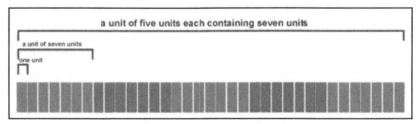

Figure 2. A structure for coordinating three levels of units.

Building on the construct of units coordinating operation and scheme (Steffe, 1992; 1994), Hackenberg and Tillema (2009) proposed three hierarchical multiplicative concepts based on how students generate and coordinate composite units. The basic criterion for distinction among the multiplicative concepts is the number of levels of units with which students operate at one time in their units coordinating activity. A student with the first multiplicative concept (MC1) can coordinate two levels of units in activity. A student with the second multiplicative concept (MC2) can take two levels of units as given and coordinate three levels of units in activity. Limitation of MC2 students is that the three-levels-of-units structure that they create in activity is not available to them in assimilation. Lastly, a student with the third multiplicative concept (MC3) can coordinate three levels of units prior to operating. Only MC3 students take three-levels-of-units structures as given and use them in assimilating problem situations. Furthermore, MC3 students can coordinate the composite units of two sequences of composite units because they are aware of each of the two sequences they are coordinating as *units*. It enables the MC3 student to flexibly switch between three-levels-of-units structures whenever necessary in the problem situations (Hackenberg, 2010; Steffe & Olive, 2010).

Hackenberg and Lee (2015) provided detailed explanations about the difference between MC2 and MC3 with an example. While engaging in the coordination of sequences of two whole numbers as seven 5s, MC3 students can maintain the structure of the result, 35, as a unit of seven units, each of which contains five units. That is, they do not lose track of the 5s as composite units as if the 5s were units of 1. In contrast, MC2 students can distribute five units into each of the seven units that they have made 35 units in all. However, in further operating 35 units become only a unit of 35 units. They do not continue to view the 35 units as a three-levels-of-units structure, such as a unit of seven units, each of which contains five units. Neither smooth nor easy is the transition from one multiplicative concept to the next as students are engaged in problem solving activity. It requires re-processing and monitoring the results of the currently used scheme by which the students can anticipate those results prior to activity. It is a significant shift in their ways of operating (Steffe, 2007).

In this chapter, we provide a fine-grained analysis of problem-solving activities of MC2 student, JuHa[1], in two different types of mathematical problem situations. Such detailed accounts of his reasoning in terms of units coordination and its generated quantitative units aided us to explain his affordability and constraint in the problem-solving activities.

---

[1] Pseudonym is used.

3 Analysis of JuHa's Problem Solving Activities

In this section, we chart JuHa's ways of solving two different types of problems (see Table 1). JuHa was one of the seven middle students in the 7th grade with whom we conducted clinical interviews. The goal of our research was to explore students' quantitative reasoning in diverse problem situations involving whole numbers and fractions, which was expected to undergird their algebraic knowledge. Many of the problems for the interview including the two problems in Table 1 were adopted from past literature (e.g. Hackenberg & Lee, 2015) whose focus had been on understanding the role of quantitative units- coordination in learning mathematics. The common aspect of the problem situations was that students were requested to carry out partitioning activities with a given figure to show their solution.

Table 1. Two types of problems.

Type	Problem
Equal Sharing	[E1] Share three identical sandwiches (shown by three identical rectangles) equally among five people. Show the share of one person in the drawing and explain how you make it. How much of a sandwich does one person get?
	[E2] The five rectangles represent five identical sandwiches. Jin-Ho wants to share the five sandwiches equally among seven people. (1) Show one person's share in the drawing and explain how you make it.
	(2) How much of a sandwich does one person eat?

Reversible Multiplicative Relationship (RMR)	[R1] The drawing below (a rectangle) represents a giant pizza that is 3 meters long. That's five times the amount of pizza Kyung-A wants to eat. Draw how much she will get. How much of a meter does she eat?

3.1 *JuHa's solving of Equal Sharing Problems*

For [E1] of equally sharing three sandwiches among five people, JuHa stared at the problem for about fifty seconds, and began to move his index finger and middle finger over the given three bars from the left end to the right of the bars. He seemed to count something using his two fingers as a measure. He put blurred lines at a point of about 2/3 of the first bars from left, 1/2 of the second bar, and 1/3 of the third bar. With a pen he pretended to count the pieces produced by his lines on the three bars (see figure 4).

Figure 4. JuHa's first partitioning for sharing three bars among five people

Then thirty seconds after he was looking at the problem, he suddenly began to partition each of the three sandwich bars into five parts and explained his partitioning activities to the interviewer.

Data Excerpt 1: Juha's distributive partitioning for equal sharing
J: Um… The easiest way is… [Laughing awkwardly] Just divide one sandwich into five and eat together. Just divide.
I: Yes. Then?
J: Yeah.
I: Then how much does one person get?

J: Um… Can it be possible to eat these [three bars] equally together by doing that?

I: You think it is not possible to eat together? [10-second pause] Why did you divide the bars like that?

J: [Pointing to the rightmost sandwich bar] This has five pieces, so eat it together. Then take another one and they eat it again.

I: How?

J: Eat again.

I: How do they eat?

J: [Pointing to the rightmost bar] These five pieces, after dividing one [sandwich]. Then eat. [Pointing to the middle bar] Again, divide this [the middle bar] into five and then five people eat them.

While musing over the problem for fifty seconds at first, he attempted to draw one person's share on the bars several times and smiled awkwardly even after he partitioned each of the three bars into five. JuHa seemed to establish a goal of equal sharing and to have a will to accomplish the goal, but the solution did not immediately occur to him. JuHa carried out counting actions with his fingers over three bars and drew three lines on the bars in an attempt to find one person's share by dividing three sandwiches into five equal parts *at once*. When he realized that it was not easy to divide the three bars into five equal parts correctly, he managed to accomplish the goal of equal sharing by putting five partitions on each bar and sharing the three five-part bars one by one among five people. That is, he could draw one person's share on the pictures by dividing the given to-be-shared quantity into several parts so that every part can be easily distributed equally among five people. His successful way was different from his previous trial in that it was performed on the assumption that equal sharing of every subset of an entire set results in equal sharing of the entire set, and the sum of equal shares of all subsets amounts to the equal share of the entire set. Especially, it was a distributive partitioning action by "sharing multiple units as multiple instances of sharing a single unit" (Wilson et al., 2011, p. 233).

For the interviewer's question of how much one person gets, he asked himself "can it be possible to eat these [three bars] equally together by doing that?" rather than finding the answer for the interview's question. It shows that he was not convinced that one person's share gained from his conducting distributive partitioning activity should be equal to the amount that he imagined while trying to divide the whole three bars into five parts at once (even though it was an estimation). It also reveals that his partitioning operation was not supported by mathematical justification[2] that a fractional part of a whole quantity equals to the sum of fractional parts of a collection of quantities consisting of the whole quantity. JuHa's partitioning operation seems to have its origin in his daily life experience of equal sharing of some foods or goods. He did not seem to expect the quantitative structures resulted from partitioning each bar into five parts prior to partitioning activity nor did he have an intent to use them. That is, he constructed a three-levels-of-units structure consisting of units of one, five and fifteen by distributing five units on each of the three bars but could not use the result of his units coordination operation as given for finding how much of a sandwich one person's share is.

As he showed uncertainty of his distributive partitioning activity satisfying the goal of equal sharing and seemed to have got lost, the interviewer encouraged him to explain how to find one person share by marking each one's share on the three five-part bars. To explicate that five people would share the three sandwiches one by one, JuHa marked each person's portion on every five-part bar with different symbols (circle, triangle, square, sector, and star) (see figure 5). The exchange followed.

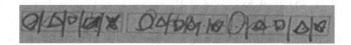

Figure 5. JuHa's marking on the bars with different symbols to distinguish five persons' shares.

---

[2] In [E1], mathematical justification involves explanation for why one person's share is 3/5 of one sandwich or 1/5 of the three sandwiches.

Data Excerpt 2: (Cont.)

J: One person has… [Using his right thumb, and left middle and ring fingers as measures alternatively, Jun counts fifteen parts of the three five-part bars by three from the right end.] One, two, three, four, five. Eat three-fifths.

I: How did you do that?

J: These [pointing to the circles on the three bars with his three fingers (left middle, ring, and little finger)] Because one, two, three, one person eats little three pieces. [Pointing to three parts of the right most five-part bar] If you… just if a person eats these [three parts] by cutting this [bar]. [Counting again 15 parts by three using his two hands] Eat three pieces, three pieces, three pieces, three pieces, three pieces. It's going to be five people.

(…)

I: [Pointing to the three parts of the right most five-part bar] You said one person eats these parts. That was your second method. Eating like that and your first method of eating a circle, a circle, a circle on the three bars, do you think they are same or different?

J: Um… Maybe they are same. They are same.

I: Are they same? You said eating like this [circle parts scattered on the three bars] and like this [three parts on a bar] are same. Right?

J: Um…the sizes are same.

JuHa reconstructed a composite unit, fifteen as structured with three levels of units with one, three, fifteen by iterating a composite unit, three five times. It seemed to make him convinced that the three parts he found should be one person's share. He was also able to tell that those parts amount to three-fifths in relation to one sandwich. It was surprising that he needed to perform counting actions using his fingers alternatively over the perceptual materials of the three five-part bars for reconstructing the whole fifteen parts as involving three levels of units. That is, he made fifteen parts (three five-part bars) through units coordinating operations that distributed a composite unit, five onto each single unit of a composite unit, three. However, as he tried to find how much the three parts are of a sandwich, the fifteen parts, the result of his partitioning operation, seemed to collapse down to a composite unit having two levels of units (one and fifteen) only. So, he had to construct a new three-levels-of-units structure for the unit, fifteen through kinetic counting activities on perceptual materials that he made before. In fact, the number, fifteen generated by his distributive partitioning activity did not seem to be a major concern to JuHa from the start. In other words, whether fifteen can be divided by five or not did not matter to him when he conducted distributive partitioning activity under the goal of equal sharing because the total number of parts of sandwiches resulted from the partitioning operation would not have any influence in achieving the goal of equally sharing the given quantity as long as he could equally share every single unit of the quantity.

For [E1], JuHa first attempted to solve the problem through what we call the '*dividing all at once* (DAO)' approach but failed. He finally drew on the given three bars how much one person gets by switching to distributive sharing. His partitioning activity was done under a '*dividing part by part* (DPP)' approach whose key idea was to get a fractional part of a whole quantity by taking separately a fractional part of a collection of sub-quantities consisting of the whole quantity and adding them up. His DPP approach, which seems to have stemmed from his daily life experience of equal sharing, was powerful enough to draw one person's share on the picture. However, it was not supported by mathematical justification, which led him to struggle to find how much one person's share was in relation to one sandwich.

For JuHa, determining how much of a sandwich the quantity that he found was a new challenging problem situation with a different goal other than equal sharing. It required re-structuring of the result of his partitioning operations conducted for equal sharing. He reconstructed a composite unit, fifteen as having a three-levels-of-units structure by iterating a composite unit, three five times and finally provided mathematical justification that the portion he found through DPP approach should be exactly one person's share.

In [E2], where five sandwiches were to be distributed equally among seven people, JuHa did not initiate partitioning activity for a while and pretended to partition over the given five sandwich bars from left end to right. It seems that he tried to divide the whole five bars into seven parts equally by visual estimation, which did not work well. First, he divided 5 by 7 using the long division method and soon realized that the answer could not be attained accurately. He turned back to the given five bars on the paper and tried to divide the five bars into seven equal parts by estimation using his fingers and a pen. In the meantime, about five minutes passed. JuHa, who was musing on the problem, suddenly divided the leftmost bar into four. The exchange followed.

Data Excerpt 3: JuHa's DPP approach for equal sharing
J: [Partitions the leftmost bar into four parts, puts lines at about 1/4 of the remaining bars, and writes "1/4" under the leftmost four-part bar.] It doesn't work. [Puts one more line on the four-part bar, which turns into a five-part bar]
I: [Pointing to the leftmost five-part bar] Why did you do like this?
J: Me? Eat four pieces, eat one [bar] at a time... Dividing into four, eat one by one...
I: Mm-hmm.
J: Wait. [Staring at the bars for fifteen seconds] Oh! [Smiles] You can divide it into thirty-five pieces.
I: Why?
J: Why? Those five bars, at first divide this [the leftmost bar] into seven. Divide into seven then every person eats one at a time. [Pointing to the second bar] This one, you can also divide it into seven, and eat one by one. Then you can keep eating like that.

JuHa's first partitioning activity, after five-minute-long struggle of dividing the whole five bars into seven, was to divide each of the five bars into four parts and distribute them to seven people part by part. Although he partitioned only one bar into four parts, based on his explanation, "eat four pieces, eat one [bar] at a time" we inferred he had mentally distributed four partitionings on the other four bars as well as the leftmost bar. JuHa soon realized the result of his partitioning activity was not suitable for equal sharing among seven people although we could not exactly tell how he had recognized such inappropriateness. One conjecture is that he mentally performed 1) distributing the first and the second four-part bars to seven people, which results in one part left, 2) distributing the third and the fourth bars, which also resulted in one part left, 3) distributing the remaining six parts (the left two parts and the fifth four-part bar), which was insufficient for seven people. Fifteen seconds later, Jun convincingly argued that the partitioning made into thirty-five parts would work. His explanation, "divide into seven then every person eats one at a time... you can keep eating like that" indicates that his solution was based on a DPP approach, which shares the whole quantity part by part at his convenience, rather than dividing the whole quantity into seven equal parts.

His claim that "you can divide it into thirty-five pieces" indicated that he had mentally partitioned each of the five sandwich bars into seven, conducting units coordinating operations of inserting seven units on each element of the five units, which resulted in a composite unit of thirty-five. However, as in [E1] we could not find any evidence he intently constructed such a three-levels-of-units structure of thirty-five with an expectation to use it for his further operation (such as taking five parts out of the thirty-five parts for one person's share). Rather, we analyzed that the number, 35, was a by-product of a DPP approach in accomplishing the goal of sharing the whole by sequential distribution of all parts in the whole. It can be called *distributive* in that equipartitioning of each unit (sandwich) by the number of people was performed for equal sharing of the whole seven units. Despite that, the use of a mathematical property of the number, 35 (for example, 35 can be divided by 7 or 35 consists of seven units of 5) did not seem to be his concern prior to the partitioning operation.

Later he displayed one person's share by circling one part on each of the five bars. As the interviewer asked how much one person eats, his answer was "5/7 pieces" because "one person eats five pieces of 1/7." The interviewer did not further probe into his answer to check whether "5/7 pieces" means 5/7 in relation to a sandwich. We had wished the interviewer had checked JuHa reorganized thirty-five parts of the five bars, which was constructed by his units coordinating operation of inserting seven units onto each of the five units (bars), into a three-levels-of-units structure of one, five and thirty-five. If he could do that, he had been able to take five parts as one-seventh of the five seven-part bars.

To summarize, JuHa initially tried to solve [E1] and [E2] with DAO approaches but could not. After switching to DPP approaches with distributive partitioning activities, he was able to find one person's share on the given pictures. In a sense, it was fortunate for JuHa to engage in distributive partitioning, that is, to set a single unit of the given quantity as to be divided for every sharing activity. Other methods based on DPP approach, for example a halving strategy[3] or dividing all single units into four as in JuHa's first method for [E2], do not guarantee equal sharing of the whole quantity only with the activity. The distributive partitioning as a strategy was a very effective method in that it enabled not only to accomplish the goal of equal sharing no matter what the number of a to-be-shared quantity and sharers were, but also to easily draw one person's share on the picture. What matters is that JuHa's distributive partitioning action can be provoked in other partitioning problems.

3.2 *JuHa's solving of RMR problem*

RMR problems are designated by Hackenberg (2010) as problems that "involve reasoning with the reverse of the stated multiplicative relationship." (p. 384) Twenty seconds after reading [R1] of the given 3/3-part bar (a three-meter pizza) being five times as much as Kyung-A wants to eat, JuHa said, "do the same for this too"[4] and partitioned the rightmost 1/3-part of the given 3/3-part bar into four subparts. Ten seconds later, he talked to himself "Ah, do I need to divide this [3/3-part bar] into five?" and hesitantly put two more lines on the rightmost 1/3-part, which came to be composed of six-subparts. Then staring at the 3/3-part bar, he divided the whole 3/3-part bar into five parts by visual estimation and erased the lines that he first drew on the rightmost 1/3-part (see Figure 6).

---

[3] For example, if a student divided each of five sandwich bars into half for seven people as in [E2], every sharer gets a half of a sandwich and three halves remain, which engenders another challenging situation.

[4] Note that [R1] was proposed to JuHa right after his solving of [E1].

Figure 6. JuHa's partitionings of the 3/3-part bar into five by visual estimation.

As JuHa pointed at the rightmost one of the five parts of the 3/3-part bar with a pen and argued Kyung-A would eat it, the interviewer asked him how much of a meter it was. He answered, "1/5 meter because three pieces are divided by five", and soon spoke to himself, "Is it wrong?" Then he started to divide 3 by 5 using the long division method and got 0.6. He told the interviewer that Kyung-A is going to eat 60 centimeters. As the interviewer asked him to solve the problem using the given 3/3-part bar rather than relying on the calculating method, JuHa, having been struggling for a while, answered he could not find any other way to solve the problem. With an additional encouragement of the interviewer to use the bar to solve the problem, he wrote "1/5" under each of the five parts of the 3/3-part bar and said, "because its' one fifth of three...one fifth times three."

As noted above, [R1] was proposed right after the equal sharing problem, [E1]. Recall that after JuHa's first attempt to use DAO approach failed, he was able to solve [E1] through activating distributive partitioning, one of the DPP approaches. However, despite that he solved [R1] right after [E1], he tried to partition a part of the 3/3-part bars into some necessary number from the beginning (i.e., a DAO approach). His behaviors of attempting to divide a part with an address, "do the same for this too" indicates he anticipated that the partitioning activities in the previous problem, [E1] somehow would be helpful in solving [R1]. Also, failing to solve [R1] by the method in the end reveals that he was not explicitly aware of why the partitioning activities were useful in [E1]. We conjecture JuHa's distributive partitioning was contextual subject to equal sharing situations in that spontaneous activation of his partitioning operations arose only in [E1] in the first interview and [E2] in the second interview. For him, it might be common to equally share all three pizzas among five people by equally sharing one by one. However, the idea of taking one-fifth of three meters by taking one-fifth of each unit of the three meters and combining them did not seem to easily occur to JuHa in the RMR situation. As his switching to divide the whole 3/3-part bar at once into five, the task was still challenging for JuHa due to his lack of splitting operation for composite units based on coordination of two three-levels-of-units structures.[5]

He managed to figure out the answer of 0.6 meter through numerical calculation. However, it did not seem to symbolize his quantitative operations with the given quantities and quantitative structures in which the operations resulted because he was confused about choosing a referent unit between a meter and the whole (three meters) in measuring the quantity that he found in the drawing. We infer that his confusion of referent units would be due to his inability to transform the 3/3-part bar into a composite unit structured with three levels of units (one, five and fifteen) and use it as given material. When he designated one-fifth of the 3/3-part bar that he drew by visual estimation as 1/5-meter, we would say, he was attending to two levels of units: a unit of the whole (three meters) and a unit of an unknown quantity, which can be iterated five times to construct the whole. However, as he was checking whether his answer, 1/5 meter, was right or wrong, a unit of the whole (three meters) has disappeared in his awareness only focusing on a unit of one meter (1/3-part) and a unit of the quantity that he drew on the 3/3-part bar. Through visual comparison of the part that he drew on the bar with one meter, he might realize that it was not at least 1/5 meter. It is one of the typical characteristics of MC2 students, attending to different two-levels-of-units structures in sequence rather than keeping aware of three levels of units simultaneously (Hackenberg & Lee, 2015). In the following conversation, the interviewer posed a question about his first partitioning of 1/3-part into six to remind him of his solving activities in [E1].

Data Excerpt 4: Jun's coordination of two three-levels-of-units structures in activity for [R1]

---

[5] Detailed discussion on the use of splitting operation with coordination of two three-levels-of-units structures is beyond this paper. For more information refer to Shin & S. J. Lee (2018) and Shin, S. J. Lee, & Steffe (2020).

I: I have a question. [Pretending to divide the rightmost 1/3-part that JuHa partitioned into six] You did like this here. Why did you do that?

J: That? [Pointing to the 1/3-part] Divide this into five. Divide all of these [three parts of 3/3-part bar] into five and then... Um... After dividing these into five, divide these into many pieces. Um... How did I do that? Um... [While the interviewer checks a video camera due to a technical issue, JuHa pretends to partition over the 3/3-part bar and wrote "15 pieces" above the 3/3-part bar. Twenty-seconds later, he suddenly drew a new rectangular bar, partitioned it into three parts, and divided each of the three parts into five subparts with writing of "3 m" above the bar.]

J: [Using his fingers of both hands alternatively as measures for counting by three] One, two, three, fourth person, fifth person. [Writing "15 pieces" above the 15-part bar] Then it's a total of fifteen pieces. Divide this one meter into five so there are three of five pieces, it's fifteen then. If you did one by one, three pieces for one person. Uh, is it wrong? What was that [what was I going to do]? It seems not... [running his fingers over the 15-part bar several times] Um... Ah! Because it is five times, [circling the rightmost three subparts of the 15-part bar with a pen] there must be five of these three to be fifteen. Then the pizza that Kyung-A wants to eat is to be these three parts. So, three parts are three-fifteenths, three meters that Kyung-A was going to eat.

I: Three-fifteenths meter?

J: Well, it might not.

I: Do not think so? Why don't you check it?

J: [Talking to himself] Is that three over fifteen pieces? [Writing "1/5" on the paper] it is one-fifth. One-fifth is, um... One-fifth meter... One-fifth meter!

I: What is one-fifth meter here [15-part bar]?

J: [Circling the rightmost three subparts of the 15-part bar] This, this one.

I: Which one?

J: [Pointing to the rightmost three parts again] this one.

In response to the interviewer's question for reminding of his first partitioning activity, JuHa seemed to try to solve [R1] by recalling his experience in [E1] of partitioning a single unit (one sandwich) into five parts. His address, "divide this into five. divide all of these [three parts of 3/3-part bar] into five" and his writing of "15 pieces" above the 3/3-part bar indicate that he constructed a composite unit, 15, by conducting units coordinating operations of inserting five units onto each part (one meter) of the 3/3-part bar without actual partitioning actions. However, it was not easy for him to transform the three-levels-of-units structure with one, five, fifteen into another three-levels-of-units structure with one, three, fifteen to take one-fifth of the whole 3/3-part bar consisting of fifteen subparts. The facts that JuHa spent twenty seconds after writing "15 pieces" and used his fingers as measures for counting fifteen subparts by three demonstrate that such transformation between two three-levels-of-units structures was not a trivial task for him. In addition, the fact that he used *person* as a unit when counting fifteen subparts by three with his fingers, and his address, "three pieces for one person" imply that he assimilated the present situation of [R1] into the context of equal sharing like [E1] in the process of recalling his partitioning activities performed in [E1]. We infer that the expression of his confusion, "Uh, is it wrong? What was that?" was a result of his conflation of the two problem situations. In the end, he did find three subparts of the three-meter bar of fifteen subparts as the portion that Kyung-A would eat by reconceiving the problem situation.

With significant prompting from the interviewer for five minutes, JuHa eventually realized that the three subparts were three-fifths of a meter. The interviewer's explicit question of how much of one meter three subparts were on the drawing of the three-meter bar composed of fifteen subparts seems to have played a critical role. He responded to the question as "there are three one-meters and divide one meter into five. Because it's three out of the five divided parts, it is three-fifths." We infer that JuHa reconceived a composite unit, fifteen as a three-level-of-units structure involving one meter as a unit of five. That is, he reconstructed the composite unit, fifteen as structured with three levels of units of one, five and fifteen. It would enable him to coordinate with another three-levels-of-units structure with one-fifth, one, and three, which has an identical multiplicative structure to the former. In fact, the structure with the units of one, five, and fifteen was what he constructed at first by partitioning each part of the 3/3-part bar into five. However, he did not consciously aware of this in the process of constructing a new three-levels-of-units structure with the units of one, three, and fifteen to take one-fifth of the 15-subpart three-meter bar. Thus, he had to recall the construction process of the structure with the units of one, five, and fifteen in identifying the three subparts as three-fifths of a meter.

In summary, JuHa, confronted with [R1], attempted to recollect his partitioning method in [E1]. However, it was an intuitive abstraction from his experience of solving process of [E1] in that he did not seem to be explicitly aware of the reason that his distributive partitioning functioned properly in [E1]. He could not take one-fifth of each part (a meter) of the given 3/3-part bar and add them together to achieve the goal of taking one-fifth of three meters. Thus, his partitioning operation was yet to be interiorized enough to activate under a non-sharing goal as in the RMR problem. Although JuHa switched to DAO approach and divided the 3/3-part bar into five at once, he had to find the answer by numeric calculation irrelevant to the quantitative operations and structures carried on the drawing of a three-meter bar. With the interviewer's prompting, he managed to partition each part of the 3/3-part bar into five subparts, but it was an imitation of his previous partitioning activities in [E1] lacking in the key idea of a distributive partitioning operation. Despite that, the result of his partitioning operation (the 3/3-part bar consisting of fifteen subparts) became a new perceptual material, which facilitated his solving [R1] with DAO approach. JuHa, by reconstructing fifteen as a composite unit consisting of five units of three, managed to mark one-fifth (three subparts) of the whole three-meter bar on the drawing. We also want to highlight that he had a hard time in finding out how much three subparts were in relation to a meter, unlike in the two equal sharing problems. We hypothesize that such difficulty might be due to the level of his multiplicative reasoning of inability to hold in mind a three-levels-of-units structure as available material for further activity. In [E1] and [E2], to measure quantities that he marked on drawing in relation to a single unit (a sandwich) might be relatively easy. In the process of conducting distributive partitioning for both problems, he established an explicit goal of sharing *each single unit* by partitioning it by the number of the sharers, which assisted him to use the single unit as a referent unit in his measuring activities. Nevertheless, considering his level of multiplicative reasoning, we suspect that he was attending only to two levels of units, for example in [E1], one-fifth and one. We have no evidence either that JuHa named three-fifteenths as one-fifth in [E1] and five thirty-fifths as one-seventh in [E2] based on the results of his equipartitioning operations for composite units.

4 Discussion

For all three problems [E1], [E2] and [R1], JuHa began by using the DAO approach. However, the DAO approach needed to accompany a splitting operation for composite units with the coordination of two three-level units structures. This is the reason why JuHa, an MC2 student, had difficulty to succeed in solving the problem using the approach.

In the equal sharing situations of [E1] and [E2], he realized soon that it was difficult for him to solve the problems with the DAO approach and managed to share the given quantity (sandwiches) by partitioning each single unit by the number of sharers and distributing them one by one. That is, JuHa's distributive partitioning activities were made possible by his spontaneous switch from the DAO to the DPP approach. However, the resulting three-level units structures (for instance, the structure with units of one, five, and fifteen by dividing each of three sandwiches into five parts in [E1]) were not what JuHa had intentionally generated to use as materials for a further step in his solving process. Even after he found one person's share with his partitioning operation, he did not seem to view the result of his activity in terms of three levels of units. In order to justify three parts should be one person's share out of fifteen parts, JuHa had to reorganize the fifteen parts of three sandwiches as a three-level units structure with one, three, and fifteen accompanied by counting activity by three using his fingers as measures.

JuHa's distributive partitioning as a DPP approach in the equal sharing situations was not generalized into the reversible multiplicative relationship problem [R1]. After he first failed with a DAO approach, he managed to obtain an answer by numerical calculation. However, he could not draw a quantity as much as the numerical answer on the given three-part bar. With the prompt of the interviewer, he recalled distributive partitioning activity that was successful in [E1], but it was a simple repetition of prior physical actions, not based on the core idea of the DPP approach that the fractional part of the whole quantity can be obtained by gathering all fractional parts of the subparts consisting of the whole.

Even after he transformed the 3/3-part of the three-meter bar into fifteen parts by putting five subparts on each part, he struggled to find one-fifth of the three-meter bar by reorganizing the fifteen subparts as a three-level units structure with one, three and fifteen. Also, it was only after a long conversation with the interviewer that JuHa told the three parts, which he found as one-fifth of the three-meter bar, should be three-fifths of a meter. We infer that these struggles of JuHa would be due to the inability to keep aware of two three-levels-of-units structures at the same time. That is, he was at best able to view the fifteen parts of the three-meter bar as a quantity only having one three-level units structure at a time.

5 Conclusion

For mathematics to be a meaningful subject in the future, as Wolfram has stressed, we must engage our students in meaningful problem solving activities connected to real life. We agree that "mathematical modeling is simply mathematics in the context of quantitative reasoning" (Thompson, 2011, p. 52). Therefore, we believe students' meaningful activities should involve quantitative operations as well as numerical calculations. Quantitative operations are not the same as numerical operations although they are related. Students often employ numerical operations that have no quantitative significance (Thompson, 2011), which is why we emphasize students' construction of quantitative units as a tool for their problem solving. In this chapter, we have demonstrated that JuHa's reasoning with two levels of quantitative units influenced how he solved the two different types of multiplicative problems. More advanced reasoning such as MC3 could open a new perspective on JuHa's understanding of the problem situations, which might lead to more successful problem solving attempts.

References

Gravemeijer, K., Stephan, M., Julie, C., Lin, F-L., & Ohtani, M. (2017). What mathematics education may prepare students for the society of the future?, *International Journal of Science and Mathematics Education, 15*(Suppl 1), 105-123.

Hackenberg, A. J. (2010). Students' reasoning with reversible multiplicative relationships. *Cognition and Instruction, 28*(4), 383–432.

Hackenberg, A. J., & Tillema, E. S. (2009). Students' fraction composition schemes. *Journal of Mathematical Behavior, 28*(1), 1–18.

Hackenberg, A. J., & Lee, M. Y. (2015). Relationships between students' fractional knowledge and equation writing. *Journal for Research in Mathematics Education, 46*(2), 196-243.

Norton, A., Boyce, S., Ulrich, C., & Phillips, N. (2015). Students' units coordination activity: a cross-sectional analysis. *Journal of Mathematical Behavior, 39*, 51-66.

Norton, A., & Boyce, S. (2015). Provoking the construction of a structure for coordinating n+1 levels of units. *Journal of Mathematical Behavior, 40*, 211-232.

Shin, J., & Lee, S. J. (2018). Conceptual analysis of students' solving equal sharing problems. In T. E. Hodges, G. J. Roy, & A. M. Tyminski. (Eds.), *Proceedings of the 40th annual meeting of the North American Chapter of the International Group for the Psychology of Mathematics Education* (pp. 215-218). Greenville, SC: University of South Carolina & Clemson University.

Shin, J., Lee, S. J., & Steffe, L. P. (2020). Problem solving activities of two middle school students with distinct levels of units coordination. *Journal of Mathematical Behavior, 59.* https://doi.org/10.1016/j.jmathb.2020.100793

Smith, J., & Thompson, P. W. (2007). Quantitative reasoning and the development of algebraic reasoning. In J. Kaput, D. Carraher, & M. Blanton (Eds.), *Algebra in the early grades* (pp. 95-132). New York: Erlbaum.

Steffe, L. P. (1992). Schemes of action and operation involving composite units. *Learning and Individual Differences, 4*(3), 259–309.

Steffe, L. P. (1994). Children's multiplying schemes. In G. Harel & J. Confrey (Eds.), *The development of multiplicative reasoning in the learning of mathematics* (pp. 3-39). Albany: State University of New York Press.

Steffe, L. P. (2007, April). *Problems in mathematics education.* Paper presented for the Senior Scholar Award of the Special Interest Group for Research in Mathematics Education (SIG-RME) at the annual conference of the American Educational Research Association in Chicago, Illinois.

Steffe, L. P., & Olive, J. (2010). *Children's fractional knowledge.* New York: Springer.

Thompson, P. W. (2011). *Quantitative reasoning and mathematical modeling.* In L. L. Hatfield, S. Chamberlain, & S. Belbase (Eds.), *New perspectives and directions for collaborative research in mathematics education.* WISDOMe Mongraphs (Vol. 1, pp. 33-57). Laramie, WY: University of Wyoming.

Wilson, P. H., Edgington, C. P., Nguyen, K. H., Pescosolido, R. C., & Confrey, J. (2011). Fractions: How to share fair. *Mathematics Teaching in the Middle School, 17*(4), 230-236.

Wolfram, C. (TED) (2000). Teaching kids real mathematics with computers. Retrieved from https://www.ted.com/talks/.

Chapter 3

Teaching Lower Secondary Statistics through the Use of Comics

Tin Lam TOH, Lu Pien CHENG
Lee Hean LIM, Kam Ming LIM

In this chapter, we discuss the comics package we have developed to teach lower secondary school statistics of the mathematics curriculum. Most people would think that comics for classroom instruction can at most stop at arousing students' interest in the subject, especially among the low-attaining students. However, we assert that using comics in teaching lower secondary statistics can introduce the students to much of the statistical processes within contexts that are meaningful to students, invite them to engage in higher order thinking tasks in order to develop their critical thinking ability. We also discuss snapshots of how one teacher enacted the statistics lessons based on our comics teaching package, and two teachers' response to the comics package.

1 Introduction

Statistics is an important component of the O-Level mathematics. In fact, Statistics and Probability forms one of the three Content Strands (together with Number and Algebra, and Geometry and Measurement) in the syllabus (Ministry of Education (MOE), 2012, p. 30).

Under the MOE curriculum document (MOE, 2012, pp. 39 - 40), the objective of statistics in the secondary school mathematics curriculum is to equip students with the tools to represent and interpret the statistical diagrams, and to make decisions using the presented data. In applying to the real world, the same document states that students should be able to read and interpret statistics in media and advertisements *critically* (emphasis added). In this chapter, we argue that using comics for classroom instruction has the advantage of meeting the above said objectives proposed by the MOE due to the nature of comics. The common assumption is that comics stops at merely arousing students' interest in the subject, especially the low-attaining students. We will show with illustration and justification that, in fact, riding on the affordance of comics, we can develop students' critical thinking skills in statistical tasks. Many of the comics related tasks can be classified as "hybrid task" (Kuntze, Aizikovitsh-Udi, & Clarke, 2017).

2 Some Preliminary Findings on Using Comics for Teaching Mathematics

We have completed one intervention research project (entitled MAGICAL) on using comics for teaching mathematics to low-attaining students (Toh, Cheng, Jiang & Lim, 2016). In that first study, we have shown that students generally responded positively to comics for mathematics instructions. We observed in the comics lessons that the participating teachers rode on the affordance of the comics instructional package to infuse various twenty-first century skills within the mathematics lessons. In the scaling up of the project MAGICAL (entitled SUPER:MAGICAL) to include more participating mainstream schools in Singapore, we observed many interesting pedagogies adopted by the various participating teachers in delivering the mathematics lessons (Toh, Chan, Cheng, Lim & Lim, 2018).

Toh, Cheng, Lim and Lim (2019) reported the result of an interview with a sample of participating teachers in MAGICAL and SUPER:MAGICAL on their opinion about using comics for teaching mathematics. The interviewed teachers appreciated the use of comics in teaching mathematics as it provides the context to scaffold students' understanding of the mathematics conveyed by the comics. Not only that, the interviewed teachers also commented that the comics provide the "real world" context, which makes mathematics relevant to their students.

There are not many research articles on using comics for mathematics instruction which are published in English language. There are few empirical studies that show that comics could raise students' motivation and interest in learning the subject (Cho, 2012; Sparrow, Kissane, & Hurst, 2010). A few other studies in English have shown that teaching mathematics using comics has the ability to reduce students' anxiety level about mathematics (e.g., Sengul & Dereli, 2010). Perhaps this quantity does not reflect the general belief of the efficacy of comics on students' learning of mathematics, as a visit to any local bookstore could point to the availability of many mathematics comics books.

In the article published by Kogo Tomoko and Kogo Chiharu (1998), it was reported in their empirical study in Japanese that presentation of instructional material using comics facilitated knowledge retention at the "shallow level"; the storyline presented in comics brought about knowledge retention required for deep learning. In their study, the learners responded that they were more interested with learning with the content presented in comics and with an accompanying storyline. It was also interesting to observe that among the research articles published on using comics for education, most of the subjects were university students. This shows that these researchers were not thinking of merely the low attainers when they used comics for their intervention and their impact on student learning.

3 Learning Statistics in School Mathematics Curriculum

Statistics is an important component of the mathematics syllabus in Singapore. This is hardly surprising as statistics has important applications for many other subjects and, more importantly, it is definitely crucial for important decision-making (e.g. Ruiz, 2004). It is not an exaggeration to say that statistics should be a common knowledge among all people, regardless of their professions.

Teaching statistics should go beyond equipping students merely with the knowledge about statistics. As Gal and Garfield (1997) aptly put it, students should be engaged in the whole process of data collection, creation and interpretation, and allowed the opportunity to bring in their prior knowledge to the learning of statistics.

Researchers have also pointed out that traditional teaching of statistics without actually engaging students in the process of statistics can likely lead to several errors such as (a) extrapolation of application contexts; (b) superficial data reading; (c) wrong choice of mode of representation, and (d) incomplete interpretation of mean, median, and standard deviation (Batanero, Godino, Green, Holmes, & Vallecillos, 2009, cited in Navarro, Delgado & Calderon, 2019).

Engaging students with the data that the students can associate with will address both the cognitive and affective factors of students learning statistics. For example, the study by Neumann, Hood and Neumann (2013) suggests that the relevance of the data is an important point for consideration in the teaching of statistics. The Statistics Research Foundation   (https://www.studentresearchfoundation.org/blog/using-student-data-to-benefit-students/) from the United States even suggested the use of student data to personalize the learning of statistics in order to ensure that statistical learning matches the students' learning needs. This is striking in the midst of the current atmosphere of strong emphasis of data protection.

4 Learning Statistics with the Help of Comics

We shall continue to use the term *comics* as an approach of conveying messages in the form of stories and which make use of imaginaries, usually in the form of cartoons (Toh, et al., 2016; 2017; 2018). The storylines used in comics are usually related to the real-world, usually an exaggeration of the real-world. The close connection of comics to the real-world usually helps students to see the relevance and fun of the message conveyed by the real world; and the use of humour has been shown to improve information retention (for example, the seminal work of Kaplan and Pascoe, 1977).

We will next describe how we tapped the above features to develop an alternative teaching package for the teaching of statistics (within the mathematics curriculum) at the lower secondary level, targeting at the lower ability students. This package has been trialled at seven secondary schools. At the time of writing this chapter, we are still in the process of analysing the data that we have collected. More of the findings from the data analysis will be reported elsewhere in the future.

4.1 *Statistics at the Lower Secondary Level*

Statistics in the Singapore school mathematics curriculum at the secondary level focuses on descriptive statistics, and inferential statistics is taught at the pre-university level. At the lower secondary level, the two main aspects in the curriculum are (1) representation of data; and (2) calculating and interpreting three measures of central tendency (mean, mode and median).

4.2 *Comics Package*

We matched all the statistical concepts which are detailed in the syllabus document (MOE, 2012). Similar to what we have done previously of the topic on percentage (Toh, et al., 2017), we designed a series of seven sets of comics for (1) representation of data; and another series of six sets of comics for (2) calculating and interpreting measures of central tendency. The objective of these sets of comics are to elicit all the concepts that are covered in the lower secondary mathematics curriculum, using the real-world context. The sets of comics involve two fictitious characters representing two diametrically opposite cognitive attribute: (1) Sam represents an observant individual who is cognizant of the environment around him, but not competent in mathematics; (2) Sarah represents a highly intellectual individual who may not be as observant as Sam. However, she is able to explain much of the real world phenomena using mathematical concepts. Through the interaction of Sam and Sarah with their surroundings, all the statistical concepts (in the school syllabuses) are elicited for discussion.

The statistics comics package consists of eight sets of comics on representation of data (Sec One) and fivesets of comics on calculation and interpretation of statistical data (Sec Two). Each set of comics was mapped to the content in the statistics section of the lower secondary mathematics syllabus as shown in Table 1.

Table 1.

Comic sets and the content

Level	Title of comics	Content
Sec 1 No. 1	Sam the Day Dreamer	Tallying and frequency table
Sec 1 No. 2	Oh My Boss!	Pictogram
Sec 1 No. 3	Employee of the Year	Bar Graph
Sec 1 No. 4	Beautiful Picture or Graph?	Line Graph
Sec 1 No. 5	Hair Tonic	Pie Chart: Calculation
Sec 1 No. 6	Support Your Idol!!!	Pie Chart: Interpretation
Sec 1 No. 7	Who is Better?	Interpretation of Statistical Diagram
Sec 1 No. 8	Pay Raise???	Misleading Statistical Diagrams
Sec 2 No. 1	MC King	Dot diagrams and Interpretation
Sec 2 No. 2	Healthy Lifestyle	Histogram and Interpretation
Sec 2 No. 3	Average Rate	Mean and from various Statistical Diagrams & Interpretation
Sec 2 No. 4	Going for Holiday	Mode and from Statistical Diagrams & Interpretation
Sec 2 No. 5	Job Recruitment Fair	Median and Consolidation of the averages: Mean, Mode & Median

Each set of the comics consists of one short story about the interaction between Sam and Sarah in their office. The story attempts to bring out the mathematical concept stated in Table 1. The related mathematical concepts were explained not using the formal expository approach, but through the dialogues of the fictitious comics characters. Figures 2 and 3 show a few windows of the comics in Sec 1 No. 1 and Sec 2 No. 3 respectively.

Figure 1. Part of Sec 1 No. 1 comics on data collection which eventually moves into the exposition of tallying and frequency table expounded by the fictitious characters.

Figure 2. Part of Sec 2 No. 3 comics that leads to the development of the concept of statistical mean through a series of lively discussion in the office situation.

4.3 *Features of the comics package*

As we intended that the comics package be all inclusive to be used as a "replacement unit" (Leong et al., 2016), we included all the proposed activities and practice questions that are used to facilitate student learning. We were mindful that we wanted the students to be engaged in the process of constructing the classroom discourse rather than be a passive comics reader. We rode on the affordance of the comics to include a series of activities, which we shall discuss below.

4.3.1 Transfer of learning through context

In designing the practice questions we moved gradually from a context identical to the comics (Figure 3), next to a similar though not identical context (Figure 4) and finally to a completely different context (Figure 5). We attempted to ensure that there would be a smooth transition across different contexts and that students would be able to transfer their learning across various contexts.

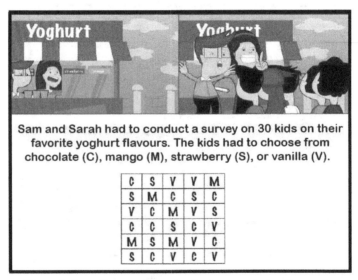

Figure 3. A practice question using the same context as the comics.

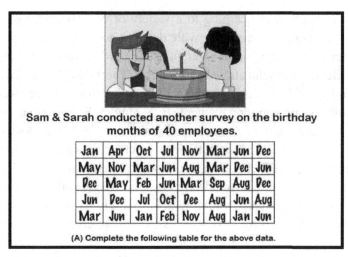

Figure 4. A practice question using a similar but not identical context, with the tabulation being slightly more complicated.

---

15 students were asked to choose which dance they prefer: ballet (B), cha cha (C), folk dance (F), hip hop dance (H), line dance (L). The results are shown below

F	L	B	L	C
B	F	L	L	F
H	L	C	F	B

(A) Complete the following table for the above data.

Dance	Number of Students
Ballet	
Cha Cha	
Folk dance	
Hip hop	
Line dance	

(B) Which was the most preferred dance?

Answer: _____

Figure 5. A practice question with a completely different context.

We believe that the transition of contexts for the practice questions was an essential factor for us to consider in order to ensure that students were able to apply their learning to a variety of contexts. This gradual transition of contexts was done with the aim to retain some similar features across two consecutive activities. According to the Theory of Identical Elements in educational psychology (Thorndike & Woodworth, 1901), transfer of learning from one activity to another is more likely if both activities share common elements (e.g. Singley & Anderson, 1989). This, in the language of Perkins and Salomon (1989), is *low road transfer*. However, including sufficient practice activities of various nature and level of difficulties in the comics package also provides the students the opportunity to mindfully abstract the general principles so that eventually the students would be able to apply to various contexts, which was termed *high road transfer*, is even more crucial for learning. In particular, for high road transfer to occur, the facilitating role of the classroom teachers in scaffolding and interacting with the students is critical (Hajian, 2019)

4.3.2 Problem posing in comics package
While it is recognized that problem solving is the heart of the Singapore mathematics curriculum, problem posing is also an important component. If problem solving is compared to the "heart" of the mathematics curriculum, then problem posing can be likened to one of the coronary vessels of the curriculum (Arikan & Unal, 2015).

Problem posing is not a recent emphasis in mathematics education. As early as 1980, the National Council of Teachers of Mathematics (NCTM) already emphasized that students must be able to solve mathematics problems in various ways and create their own problems under various situations (NCTM, 1980). The Australian Education Council (1991) also showed great support for the use of open-ended problems in mathematics classrooms. According to the document, "[s]tudents should engage in extended mathematical activities which encourage problem posing, divergent thinking, reflection and persistence..." (p. 39).

Anecdotal evidence from the Singapore mathematics classrooms shows that students usually do not have the opportunity to be engaged in the processes of problem posing. This is commonly attributed to the lack of curriculum time and a relatively packed syllabus. For the low-attaining students, it is also a common belief among teachers that the weaker students are not able to perform problem posing.

In our design of the package, we attempted to introduce elementary problem posing tasks in structuring our practice questions. Elementary problem posing tasks were introduced as part of the questions on interpreting the statistical diagrams (Figure 6).

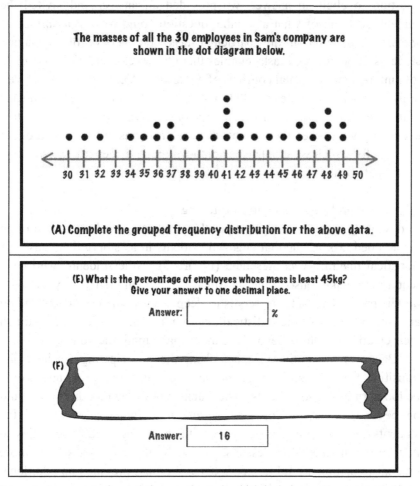

Figure 6. An excerpt of a statistics practice task which includes an elementary problem posing task as the last part of the activity.

The activity in Figure 6, from Sec 1 Set 6 (Support Your Idol!!!) includes engaging the students in three main activities: (1) representating a set of data using a frequency table and histogram; (2) interpreting the data through answering five parts of scaffolding questions (not shown in Figure 6); and (3) posing problem with a given answer.

We attempted to introduce the open-ended task (F) (See Figure 6) by engaging students in problem posing. Instead of asking the students a question to elicit an answer, we provided an answer and invited the students to think of what a suitable question could be *in relation to the statistical diagram* of this particular task. We believe that engaging students in open-ended tasks enables them to develop creativity for them to think beyond the usual confines of a question (Yeap, 2008). However, unlike the activity of Yeap (2008, p. 46), we did not provide the students with the full-scale open-ended response. We were mindful that eventually we wanted the students to pose problems in the context of the activity, for which we emphasized on developing a deeper understanding of the statistical diagrams.

4.3.3 From problem posing to storytelling
In designing the comics teaching package, we extended the notion of an open-ended task to one that engaged students in a "storytelling" task. A statistical diagram was presented (Figure 7) to the students, who were required to make sense of the statistical diagram and tell a story based on the diagram. This activity serves to deepen the students' understanding and interpretation of the statistical diagram and, at the same time, stretch their creativity to think beyond the usual mathematics lessons.

Based on an initial feedback from the participating teachers, the transition from the problem posing tasks to this storytelling task would be too abrupt for the students. The teachers believed that this task would be beyond the ability of the low-attaining students. Moreover, the students had not much experience with such open-ended activities in their usual mathematics lessons. As such, it was suggested a more gradual transition would be necessary. We next provided an additional scaffold for the students (Figure 8).

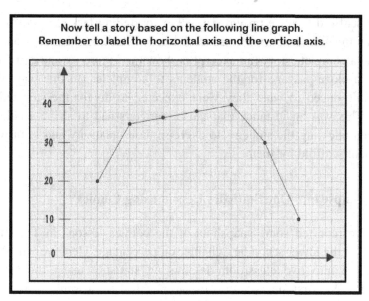

Figure 7. An open-ended storytelling task based on a line graph.

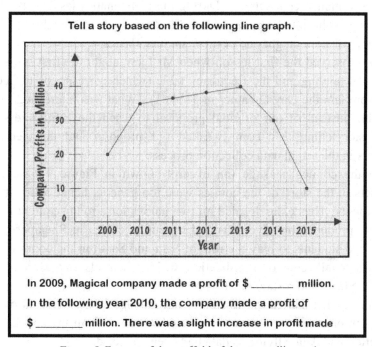

Figure 8. Excerpt of the scaffold of the storytelling task.

In the scaffold that we provided for the storytelling task (Figure 8), the axes of the line graph were labelled for the students. In addition, we created a complete story (an excerpt is shown in Figure 8), and converted it into a cloze passage for the students to complete. In this way, they were introduced to what they were expected to do for the storytelling task. This was an instance that the researchers had to strike a balance between allowing the students to exercise their creativity and to provide sufficient scaffold for them.

5 A Glimpse into One Statistics Lesson using Comics

We have reported some snapshots of the comics lessons in Toh et al. (2017) on the lower secondary chapter on percentage. Here we will show a glimpse into two classroom lessons on statistics using the comics package. We first discuss the lesson of Mr Lam (pseudonym), who taught half a class of Secondary One Normal Technical students in a Singapore mainstream school. During the mathematics lessons, the full class was split into two smaller classes which were taught by two teachers. Mr Lam taught one of these smaller classes.

We observed the first lesson when Mr Lam taught the first statistics lesson on tallying and the frequency table (which corresponds to Sec 1 No. 1 on Sam the Day Dreamer in Table 1). What was interesting to us was that instead of our original proposal to teach the comics lessons using storytelling, Mr Lam tweaked our proposal and engaged his students in the actual data collection process.

In the comics package (an excerpt shown in Figure 9), the five products in the story of the comics were too generic, hence abstract for his students. Hence, he decided on two innovations: to (1) contextualize the five products into five fast-food restaurant chains in Singapore; (2) select two students to play the role of Sam and Sarah in the comics, and to do a small-scale data collection within their classmates on their favourite fast-food restaurant (Figure 9).

The teacher started the lesson by introducing the background of the story, but engaged the two role play students to perform the actual data collection and data tallying as data had been collected.

Figure 9. The five products were contextualized as five fast-food restaurant chains in Singapore and two students were selected to play the role of Sam and Sarah in the lesson.

Figure 10. The students were performing data tallying on the board and discussing how the tallying could be done.

As shown in Figure 10, the two role-play students were hesitating on how to tabulate the data when the frequency for one category became very large. A series of lively discussion went on in this part of the lesson, before the teacher intervened. After this whole data collection process, the teacher got the students to read the whole comics and summarized the key learning points in this lesson.

Mr Lam aptly adapted the statistics comics package by contextualizing the comics to one that his students were most familiar with, and made use of the storyline of the comics to engage his students in the actual data collection and data representation process. Instead of providing the usual exposition on how tallying could be best done, he engaged his students to attempt the tallying processes itself and use their common sense knowledge to appreciate the most efficient way of tallying. In this entire process, we observed that he managed to clarify students' misconception in the process of tallying.

6 Two Teachers' Responses on the Statistics Comics Package

We conducted interview with two participating teachers, Mr Han and Ms Sun (pseudonyms), from two mainstream schools who reviewed the package after they had conducted the statistics lessons using the comics package. Their feedback about the statistics comics package can be classified under two categories: (1) context of the comics; and (2) practice questions.

6.1 *Story and context of the comics*

Both teachers thought that the stories of the comics helped to engage the students during lessons. Both Mr Han and Ms Sun were relatively new teachers with less than five years of teaching experience at their respective schools. Due to administrative constraint, we were not able to conduct an interview with Mr Lam.

Mr Han commented that this was the first time he could get his Normal Technical students to "discuss mathematics" during the lessons. For the earlier lessons, he would need to spend much time in maintaining discipline during the mathematics lessons. In particular, he highlighted that the humour in the comics within the storyline captured his students' attention during the lessons.

Ms Sun commented that although she and her students appreciated the comics and the context of the comics, she commented that occasionally the context might not be too relevant to her students (e.g.

pay raise, events at the office). She commented that occasionally, she needed to provide the contextual knowledge of the adult world before the story could make sense to her students. She compared the context of the comics of the statistics package to that of the percentage package (the first comics package that we developed earlier, reported in Toh et al. (2017)), and commented that the context of the latter package, which involves the shopping encounters and the hilarious incidents that they saw during their shopping trip, was more relevant to her students.

6.2 *Problem Posing and the Storytelling Tasks*

Mr Han commented that his students responded well to the problem posing tasks, although they might not have encountered much of problem posing tasks in the usual mathematics lessons. Mr Han in particular, administered the story-telling task (Figure 7) to his students. However, it appears that several of his students were not clear about what they were expected to do, and in fact confused the story-telling task with a problem posing task. Instead of narrating a story based on the graph, the student posed questions based on the graph. A sample of the student's work is shown in Figure 11.

Another observation of the student work in Figure 11 was that the students had used categorical data in labelling the horizontal axis, which is inappropriate since the categories were not even ordinal. The teacher could have built on this moment to discuss with his students about the nature of data, and the objective of data representation using a line graph.

Ms Sun commented that her students had not been exposed to problem posing in mathematics lessons. Further, this type of problem posing tasks is not common in examinations. Hence, she had chosen to skip these higher order tasks. However, she recognized the importance of these within the larger framework of mathematical problem solving in the mathematics curriculum. Perhaps more discussion between the teachers and the researchers could prepare them in enacting such tasks in the mathematics classroom.

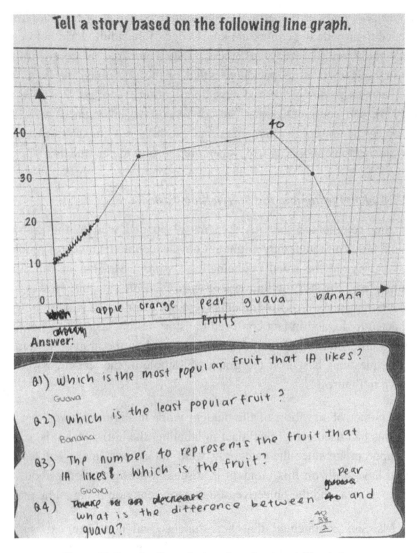

Figure 11. Sample of a student work on the story telling task.

7 Conclusion

In the development process and in consultation with much mathematics education literature, we believe that using comics for teaching secondary school statistics has much potential for *all* students, rather than restricted to the low attainers. Based on our discussion above, it is clear that comics for instruction in statistics can possibly stretch students to appreciate statistics and mathematics in the real world. The problem remains at the enactment phase. Perhaps more of teacher professional development and teacher engagement in the research process itself could close the gap between the theory and the practice.

Acknowledgement

This chapter is based on the research project No. AFR04/16TTL funded by Singapore Ministry of Education (MOE) under the Education Research Funding Programme and administered by National Institute of Education (NIE), Nanyang Technological University, Singapore. Any opinions, findings, and conclusions or recommendations expressed in this material are those of the author(s) and do not necessarily reflect the views of the Singapore MOE and NIE.

References

Australian Education Council (1991). A national statement on mathematics in Australian Schools. Canberra: Author.

Arikan, E. E., & Unal, H. (2015). Investigation of problem-solving and problem-posing abilities of seventh-grade students. *Education Sciences: Theory and Practice, 15* (5), 1403 – 1416.

Batanero, C., Godino, J., Green, D., Holmes, P., & Vallecillos, A. (2009). Errores y dificultades en la comprensión de los conceptos estadísticos elementales. *Internation Journal of Mathematics Education in Science and Technology, 25*(4), 527-547.

Cho, H. (2012). *The use cartoons as teaching a tool in middle school mathematics.* Seoul: ProQuest, UMI Dissertations Publishing.

Gal, I., & Garfield, J. B. (1997). Curricular goals and assessment challenges in statistics education. In I. Gal, & J. B. Garfield (Eds.), *The assessment challenge in statistics education* (pp. 1 – 13). Amsterdam: The International Statistics Institute.

Hajian, S. (2019). Transfer of learning and teaching: A review of transfer theories and effective instructional practices. *IAFOR Journal of Education, 7* (1), 93 – 111.

Kaplan, M., & Pascoe, C. (1977). Humorous lectures and humorous examples: Some effects upon comprehension and retention. *Journal of Educational Psychology, 69*, 61–65.

Kogo Tomoko and Kogo Chiharu (1998), マンガによる表現が学習内容の理解と保持に及ぼす効果. 日本教育工学会論文誌 / 日本教育工学雑誌, *22*(2), 87–94.

Kuntze, S., Aizikovitsh-Udi, E. & Clarke, D. (2017). Hybrid task design: Connecting learning opportunities related to critical thinking and statistical thinking. *ZDM Mathematics Education, 49*: 923 – 935.

Leong, Y.H., Tay, E.G., Toh, T.L., Yap, R.A.S, Toh, P.C., Quek, K.S., & Dindyal, J. (2016). Boundary objects within a replacement unit strategy for mathematics teacher development. In B. Kaur, O.N. Kwon, & Y.H. Leong (Eds.), Professional Development of Mathematics Teachers: An Asian Perspective (pp. 189-208). Singapore: Springer.

Ministry of Education Singapore (MOE). (2012). *O-level mathematics teaching and learning syllabus.* Singapore: Ministry of Education.

National Council of Teachers of Mathematics. (1980). An agenda for action: Recommendations for school mathematics of the 1980s. Resron, VA: Author.

Neumann, D. L., Hood, M., & Neumann, M. M. (2013). Using real-life data when teaching statistics: student perceptions of this strategy in an introductory statistics course. *Statistics Education Research Journal, 12* (2), 59 – 70.

Navarro, C., Delgado, I., & Calderon, M. G. (2019). Multimedia instructional unit for the approach of statistical topics in the high school diploma for adults program using the eXeLearning technological tool. *Propósitos y Representaciones, 7*(2), 75 – 106.

Perkins, D. N., & Salomon, G. (1989). Are cognitive skills context-based? *Educational Researcher, 18*(1), 16 – 25.

Ruiz Muñoz, David. (2004). Manual de Estadística. Universidad de Málaga. Mediterraneam Network.
Retrieved from: http://www.eumed.net/cursecon/libreria/drm/drm-estad.pdf

Sengul, S., &. Dereli, S. (2010). Does instruction of "Integers" subject with cartoons effect students' mathematics anxiety? *Procedia – Social and Behavioral Sciences, 2*, 2176–2180.

Singley, M. K., & Anderson, J. R. (1989). *The transfer of cognitive skill* (No. 9). Harvard University Press.

Sparrow, L., Kissane, B., & Hurst C. (Eds.). (2010). *Shaping the future of mathematics education: Proceedings of the 33rd annual conference of the Mathematics Education Research Group of Australasia* (pp. 515–522). Fremantle: MERGA.

Thorndike, E.L., & Woodworth, R.S. (1901). The influence of improvement in one mental function upon the efficiency of other functions. *Psychological Review, 8,* 247-261.

Toh, T.L., Chan, C.M.E., Cheng, L.P., Lim, K.M., Lim, L.H. (2018). Use of Comics and Its Adaptation in the Mathematics Classroom. In Toh, P.C., Chua, B.L. (Ed.), *Mathematics Instruction: Goals, Tasks and Activities* (pp. 67 - 86). Singapore: World Scientific.

Toh, T.L., Cheng, L.P., Ho, S. Y., Jiang, H., & Lim, K. M. (2017). Use of comics to enhance students' learning for the development of the twenty-first century competencies in the mathematics classroom. *Asia-Pacific Journal of Education, 37* (4), 437 – 452.

Toh, T.L., Cheng, L. P., Jiang, H., & Lim, K. M. (2016). Use of comics and storytelling in teaching mathematics. In P. C. Toh, & B. Kaur (Eds.), *Developing 21st Century Competencies in the Mathematics Classroom, Yearbook 2016, Association of Mathematics Educators* (pp. 241-260). Singapore: World Scientific Publishing.

Toh, T.L., Cheng, L.P., Lim, L.H., & Lim, K.M. (2019). Shopaholics need mathematics too! Teacher and student perception of the use of comics to teach mathematics. *Australian Mathematics Education Journal, 2019*(1), 1-15.

Yeap, B. H. (2008). Teaching of algebra. In P. Y. Lee (Ed.), *Teaching secondary school mathematics: A resource book* (pp. 25 – 50). Singapore: McGraw-Hill Publishing.

Chapter 4

Models of Instruction and Mathematics Teaching in Classrooms of Singapore Secondary Schools

Berinderjeet KAUR, Cherng Luen TONG

A model of instruction is a set of strategies that guides teachers in their instructional practice. The purpose of this chapter is to dispel the myth that mathematics teaching in Singapore schools is all about drill and practice, as perceived of many Asian systems. This chapter draws on data of a large project that examined the enactment of school mathematics curriculum in Singapore secondary schools. Based on the teaching practices of 30 competent teachers, a survey was constructed and administered to 677 teachers. The data from the survey showed that teachers go well beyond traditional forms of instruction in their teaching practices in Singapore secondary schools.

1 Introduction

Leung (2001) noted that in East Asian mathematics classrooms:

> Instruction is very much teacher dominated and student involvement minimal. ... [Teaching is] usually conducted in whole group settings, with relatively large class sizes. ... [There is] virtually no group work or activities, and memorization of mathematics is stressed ... [and] students are required to learn by rote. ... [Students are] required to engage in ample practice of mathematical skills, mostly without thorough understanding. (Leung, 2001, pp. 35–36).

Hogan et al (2013), as part of the CORE 2 study in Singapore carried out by the Office of Education at the National Institute of Education, examined the instructional practices of Grade 9 mathematics teachers and found that several models of instruction were prevalent in the practices. They claimed that all of these models had the goal of mastery and examination preparation. In a synthesis of past mathematics classroom studies done in Singapore, Kaur (2017) conjectured that instructional practices for mathematics in Singapore classrooms, based on the data of the CORE 2 study and the Learner's Perspective Study (LPS) carried out in Singapore (Kaur, 2009), cannot be considered either Eastern or Western but a coherent combination of both. The basis of the claim is that: i) Traditional Instruction (TI) provides the foundation of the instructional order, and ii) Direct Instruction (DI) builds on TI practices and extends and refines the instructional repertoire. Teaching for Understanding/ Co-regulated Learning Strategies (TfU/CRLS) practices build on TI and DI practices and extend the instructional repertoire even further in ways that focus on developing student understanding and student-directed learning. In this chapter, we will describe TI, DI, TfU and CRLS as models of instruction and the Enactment Project in greater details to further illuminate models of teaching practices of mathematics teachers in Singapore secondary schools.

2 Models of Instruction

A model of instruction is a set of strategies that guides teachers in their instructional practice. Often, such models are based on learning theories. Four main models are as follows.

2.1 *Traditional Instruction*

A method of instruction that is teacher-centred, rather than learner-centred, in which the focus is on rote-learning and memorization. In the context of Asian classrooms it is often associated with drill and practice

(Hogan et al., 2013; Biggs & Watkins, 2001; Leung, 2006). Examples of teacher action that depict this model are:

- Using the textbook to introduce concepts / skills.
- Emphasizing basic facts/steps and asking students to memorize them.
- Telling students how to do it when they face difficulty with a mathematical task.

2.2 *Direct Instruction*

A method of instruction that involves an explicit step-by-step strategy, often teacher-centred, with checks for mastery of procedural or conceptual knowledge (Hogan et al., 2013; Hattie, 2003; Good & Brophy, 2003). Examples of teacher actions that depict this model are:

- Using exposition (teacher at the front of class talking to the whole class) to explain mathematical ideas, facts, generalizations.
- Demonstrating how to apply a concept/carry out a skill – demonstrating again the same using a similar example but with inputs from students – asking the students to do a similar question by themselves.

Examples of student actions that depict this model are:

- Asking questions when they do not understand.
- Practising a similar problem after teacher has shown them how to do it on the board.

2.3 *Teaching for Understanding*

A method of instruction that places student learning at the core. Teacher facilitates, monitors and regulates student learning through student-centred approaches (Hogan et al., 2013; Perkins, 1993; Good & Brophy, 2003). Examples of teacher actions that depict this model are:

- Focusing on mathematical vocabulary to help students build mathematical concepts / adopt the correct skills needed to work on mathematical tasks.

- Asking students open-ended questions and allowing them to build on each other's responses to develop concepts or clarify their understanding.

Examples of student actions that depict this model are:

- Developing concepts through exploratory/investigative activities.
- Analysing why a procedure (that teacher has shown on the board) works or why a solution method makes sense.

2.4 *Co-Regulated Learning Strategies*

This model essentially involves self-directed learning, self-assessment and peer-assessment (Hogan et al., 2013; Black et al, 2003; Wiliam, 2007). Examples of teacher actions that depict this model are:

- Getting students to set their own learning goals for mathematics at the beginning of each school term/semester.
- Getting students to grade their own mathematics work (with the marking scheme/rubric provided and teaching them how to use it).

Examples of student actions that depict this model are:

- Reviewing their mistakes and identifying possible causes by themselves.
- Exploring alternative solution methods for a problem besides the one the teacher has shown on the board.

3 The Enactment Project

The project studied the interactions between secondary school mathematics teachers and their students, as it is these interactions that fundamentally determine the *nature* of the actual mathematics learning and teaching that take place in the classroom. As part of the project, we examined the content through the instructional materials used—the preparation of these materials and their use in mathematics teaching and learning. Such studies are crucial for the Ministry of Education (MOE) in Singapore and schools to gain a better understanding of what works in the instructional core in their classrooms and schools. This is also critical for the development of the education system in Singapore.

The project has two phases: the video study and the survey study, which was dependent on the findings of the video phase. The video study documented the pedagogy of competent experienced secondary mathematics teachers while the survey segment aimed to establish how prevalent the pedagogy of these competent experienced teachers is in the mathematics classrooms of Singapore schools. The video segment of the study adopted the complementary accounts methodology developed by Clarke (1998 & 2001), a methodology which is widely used in the study of classrooms across many countries in the world as part of the Learner's Perspective Study (Clarke, Keitel, & Shimizu, 2006). This methodology recognizes that only by seeing classroom situations from the perspectives of all participants (teachers and students) can we come to an understanding of the motivations and meanings that underlie their participation. It also facilitates practice-oriented analysis of learning. For the survey, the project adopted a self-report questionnaire to collect data on teachers' enactment of their "teacher-intended" curriculum.

Thirty competent experienced teachers (10 Express course of study, 4 Integrated Programme, 8 Normal (Academic) course of study and 8 Normal (Technical) course of study) and 447 (about 8 to 20 students from each class, who volunteered to be the focus students, were interviewed) students in their classrooms participated in the video segment of the project. In the context of the project, a competent experienced teacher is one who has taught the same course of study for a minimum of 5 years, and is recognized by the school / cluster as a competent experienced teacher who has developed an effective approach of teaching mathematics. These teachers were nominated by their respective school leaders and the research team followed up on the nominations and interviewed the teachers. A strict requirement for participation in the study was that the teacher had to teach the way she / he did all the time, i.e., no special preparation was expected.

For the survey segment of the project, 689 secondary school mathematics teachers from 109 schools in Singapore participated in the project. In the preliminary screening of the data, some responses were omitted as they did not meet the requirements of the survey, such as a minimum of three years of mathematics teaching practice in the course of study for which they opted to do the survey. The data of 677 teachers were analyzed and part of it reported in this chapter.

3 The Survey

The survey was based on the findings in phase 1. It was an online survey comprising three sections: pedagogical structure and student-teacher interaction (60 items); enactment of the different facets of the "pentagon framework" (MOE, 2012) undergirding the secondary mathematics curriculum (78 items); and instructional materials (226 items, of which a participant needs only respond according to one of the subjects: Additional Mathematics, Elementary Mathematics or Normal Academic Mathematics). In this chapter, we focus on the first section of the survey.

3.1 *Items and participants*

The 60 items in the first section were divided into two parts of 36 and 24 items respectively. Every item was linked to a form and purpose of *classroom talk* (rote, recite, instruct, expose, discuss, scaffold dialogue, narrate, explain, analyse, speculate, explore, evaluate, argue, justify), a *phase of instruction* (development, seatwork or review) and a *model of instruction* (traditional instruction, direct instruction, teaching for understanding or co-regulated learning strategies).

The first part had items which elicited responses about what the teacher did in class. An example of such an item is as follows:

- I ask students to recall past knowledge.

  [*rote, review, traditional instruction*]

The second part had items which elicited responses about what the teacher wanted the student to do in class. An example of such an item is as follows:

- I get my students to explain how their solutions or how their answers are obtained.

  [*explain, development, teaching for understanding*]

The survey respondents were Singapore secondary school teachers from four different academic courses: Integrated Programme (IP) (58),

Express (EX) (380), Normal Academic (NA) (151), and Normal Technical (NT) (88). These four academic courses are broadly based on academic achievement of students in the Primary School Leaving Examination. The most academically inclined students are in the IP and the least inclined ones are in the NT course of study. The on-line survey was administered to teachers following their written consent to participate in the project.

Participants were asked to reflect on their lessons for a course (IP, EX, NA or NT) they were teaching and respond to the items indicating the frequency of their actions on a Likert Scale of 1 (Never/Rarely) to 4 (Always). Of interest to us in this chapter are 57 (7 for TI, 11 for DI, 28 for TfU and 11 for CRLS) items from the first part of the survey that focuses on teacher and student actions for the four models of instruction, traditional, direct, teaching for understanding and co-regulated learning strategies.

3.2 *Data and findings from the survey*

3.2.1 A hybrid model (TI + DI + TfU)

Table 1 shows the means for three models of instruction, namely TI, DI and TfU, for all the teachers according to the 4 courses of study (IP, EX, NA, NT). These three models of instruction were apparent in the lessons of the 30 experienced and competent teachers who participated in the first part of the project.

Table 1

Mean for each of the models of instruction

Course of Study	Mean[+] Model of Instruction		
	TI	DI	TfU
All (n=677)	2.78	3.11	2.86
IP (n=58)	2.42	3.07	3.00
EX (n=380)	2.78	3.10	2.88
NA (n=151)	2.81	3.10	2.77
NT (n=88)	2.94	3.17	2.85

[+]maximum = 4; minimum = 1.

Table 1 shows that teachers appear to draw on teaching moves from all the three models of instruction with differing emphasis in their enactment of lessons. Direct Instruction appear to be the dominant model that teachers draw on in all the four courses of study. In the NA and NT classes, Direct Instruction and Traditional Instruction are apparently more prevalent whilst in the IP and EX classes Direct Instruction and Teaching for Understanding are apparently more prevalent. We next examine the survey items for each course of study that had a mean greater than 3 and a standard deviation of less than or equal to 0.7. Tables 2, 3 and 4 show teacher / student actions from the three models of instruction that were frequent in the classrooms of the teachers who participated in the survey.

Table 2

Traditional Instruction actions that were frequent in mathematics lessons

Item	IP	EX	NA	NT
Teacher asking direct questions to stimulate students' recall of past knowledge / check for understanding of concepts being developed in the lesson.		*	*	*
Teacher providing students with sufficient questions from textbooks / workbooks / other sources to practise so as to develop procedural fluency.	*	*	*	*
Teacher getting students to automatize steps leading to a solution through repetitive exercises.				*

Table 3

Direct Instruction actions that were frequent in mathematics lessons

Item	IP	EX	NA	NT
Teacher using the "I do, We do, You do" strategy, i.e. 1. Demonstrating how to apply a concept / carry out a skill on the board [I do]. 2. Demonstrating again the same using another similar example but with inputs from students [We do]. 3. Asking the students to do a similar question by themselves [You do].		*	*	*
Students practising a similar problem after teacher has shown them how to do it on the board.		*	*	*
Students asking questions when they do not understand.	*	*	*	*
Teacher walking around the class and providing students	*	*	*	*

with between desk instruction (i.e. help them with their difficulties) when they are doing their work at their desks.				
Teacher walking around the class noting student work that teacher would draw on to provide the class feedback during whole class review.	*	*	*	*
Teacher using exposition (Teacher at the front talking to whole class) to explain mathematical ideas, facts, generalisations.		*		
Teacher using a set of teacher customised notes cum worksheets to introduce concepts/skill.	*			
Teacher using a set of teacher customised notes cum worksheets to engage students in practice (application of concepts).	*			
Teacher explaining what exemplary solutions of mathematics problems must contain (logical steps and clear statements and / or how marks are given for such work during examinations).	*	*	*	
Teacher only progressing to the next objective of the lesson when he/she is confident that students have grasped the one before.	*	*	*	*

Table 4

Teaching for Understanding actions that were frequent in mathematics lessons

Item	IP	EX	NA	NT
Teacher providing feedback to individuals for in-class work and homework to serve as information and diagnosis so that students can correct their errors and improve.	*	*	*	*
Teacher providing collective feedback to whole class for common mistakes and misconceptions related to in-class work and homework.	*	*	*	*
Teacher focusing on mathematical vocabulary (such as factorise, solve) to help students adopt the correct skills needed to work on mathematical tasks.	*	*	*	*
Teacher focussing on mathematical vocabulary (such as equations, expressions) to help students build mathematical concepts.	*	*	*	
Teacher asking questions to encourage reasoning and speculation, not just elicit right answers.	*	*	*	
Teacher using examples and non-examples to engage students in discussion to make sense of a concept.	*			
Teacher leading whole class discussion (with guided questions) to facilitate the development of concepts.	*			
Teacher building on students' responses rather than	*			

merely receiving them.				
Students predicting ideas and justifying them (Example: What is a zero vector? Has it got a magnitude and a direction?).	*			
Students asking questions (such as 'what if') to probe further or for deeper understanding.	*			
Teacher providing students with probing guidance (open-ended questions about their thinking and why they are considering certain approaches) when they face difficulty with a mathematical task they are doing.	*	*		
Teacher focusing on mathematical processes (such as compare and contrast, logical reasoning), to facilitate the development of concepts or student understanding.	*	*		
Students analysing why a procedure (that teacher has shown on the board) works or why a solution method makes sense.	*	*		
Students explaining how their solutions or how their answers are obtained.	*	*	*	*

From Tables 2, 3 and 4, it is apparent that teachers from all four courses of study drew on a multitude of teaching actions from the three models when enacting their lessons. It appears that, though ability of students may be a limiting factor at times for teacher actions such as "teacher using examples and non-examples to engage students in discussion to make sense of a concept", there are several teacher actions that are consistent in our mathematics lessons irrespective of student abilities. These are:

- Teacher providing students with sufficient questions from textbooks / workbooks / other sources to practise so as to develop procedural fluency.
- Students asking questions when they do not understand.
- Teacher walking around the class and providing students with between desk instruction (i.e. help them with their difficulties) when they are doing their work at their desks.
- Teacher walking around the class noting student work that teacher would draw on to provide the class feedback during whole class review.
- Teacher only progressing to the next objective of the lesson when he/she is confident that students have grasped the one before.

- Teacher providing feedback to individuals for in-class work and homework to serve as information and diagnosis so that students can correct their errors and improve.
- Teacher providing collective feedback to whole class for common mistakes and misconceptions related to in-class work and homework.
- Teacher focusing on mathematical vocabulary (such as factorise, solve) to help students adopt the correct skills needed to work on mathematical tasks.
- Students explaining how their solutions or how their answers are obtained.

3.2.2 Empower learners to take charge of their learning

There was little evidence from the video-recorded lessons about the model of instruction: Co-Regulated Learning Strategies (CRLS), which basically involves self-directed learning, self-assessment and peer-assessment (Hogan et al., 2013; Black et al, 2003; Wiliam, 2007). The only student actions that were occasionally present in the lessons of the 30 experienced and competent teachers were students explaining how to correct an error or a misconception that the teacher put on the board and exploring alternative solution methods for a problem besides the one the teacher had shown on the board. Nevertheless, we included teacher/student actions that belonged to this model of instruction in the survey to explore how teachers in general would respond to them. Examples of actions that belonged to this model of instruction were:
Teachers getting students to:

- set their own learning goals for mathematics at the beginning of each school term/semester.
- make a plan to revise their work and correct their mistakes.
- getting students to work with peers to make a plan for revision and correction of mistakes.
- grade their own mathematics work (with the marking scheme / rubric provided and teaching them how to use it).
- identify strategies that would help them achieve their learning goals for mathematics.

- show him/her their work and review their progress for mathematics.

Students to:

- explain how to correct an error or a misconception that teacher has put on the board.
- review their mistakes and identify possible causes by themselves.
- explore alternative solution methods for a problem besides the one the teacher has shown on the board.
- work with peers to review their mistakes, identify and justify possible causes.
- ask self/classmate questions to check their understanding.

Table 5 shows the mean for CRLS model of instruction for all teachers. From Table 5 it is apparent that on average teachers in all four courses of study do sometimes guide students in self-directed learning and engage them in self and peer assessment.

Table 5

Means for CRLS model

Course of Study	Mean[+]
All (n=677)	2.54
IP (n=58)	2.54
EX (n=380)	2.57
NA (n=151)	2.46
NT (n=88)	2.58

[+]maximum = 4, minimum = 1

4 Discussion and Concluding Remarks

The data presented in this chapter allows us to gain the following insights about the teaching and learning of mathematics in Singapore secondary schools. Firstly, it is apparent that teachers do not draw on any one model of instruction when enacting their mathematics lessons. They draw on a few models of instruction, namely direct Traditional Instruction, Direct Instruction and Teaching for Understanding. This affirms a hybrid model comprising elements of Traditional Instruction, Direct Instruction and Teaching for Understanding, as speculated by Hogan and his colleagues (2013). It also lends support to Kaur's (2017) conjecture that instructional practices for mathematics in Singapore classrooms are neither Eastern nor Western but a combination of both. The teaching and learning of mathematics in Singapore secondary school go well beyond drill and practice, a stereotype of Asian Mathematics classrooms.

Secondly, Traditional and Direct Instructions provide students with mainly teacher-centred learning experiences. For students to also engage with student-centred learning and be empowered to take ownership of their learning, Teaching for Understanding and Co-Regulated Learning Strategies teacher/student actions must also pervade the instructional practices of teachers in Singapore secondary school mathematics classrooms. To do this, teachers need to shift their roles from being mainly a disseminator of knowledge to a facilitator of knowledge construction. In addition, they must also believe that their students are capable of taking charge of their own learning. To this end, they must provide students with opportunities engage with self-directed learning, self and peer assessment.

Lastly, as made apparent by the OECD Learning Framework 2030 (OECD, 2019), students must learn to navigate by themselves in unfamiliar contexts and find their direction in a meaningful and responsible way, instead of simply receiving fixed instructions or directions from their teachers. This reinforces the dire need for teachers to empower their students to take charge of their mathematics learning in Singapore secondary schools.

Acknowledgement

This chapter is based on the Programmatic Research Project: A Study of the Enacted School Mathematics Curriculum funded by Singapore Ministry of Education (MOE) under the Education Research Funding Programme (OER 31/15 BK) and administered by National Institute of Education (NIE), Nanyang Technological University, Singapore. Any opinions, findings, and conclusions or recommendations expressed in this material are those of the author(s) and do not necessarily reflect the views of the Singapore MOE and NIE.

References

Biggs, J. & Walkins, D. (2001). *Teaching the Chinese learner: Psychological and pedagogical perspectives*. Hong Kong: The University of Hong Kong, Comparative Education Research Centre.

Black, P., Harrison, C., Lee, C., Marshall, B. & Wiliam, D. (2003). *Assessment for learning: Putting it into practice*. Milton Keynes: Open University Press.

Clarke, D. J. (1998). Studying the classroom negotiation of meaning: Complementary accounts methodology. In A. Teppo (Ed.), *Qualitative research methods in mathematics education: Journal for Research in Mathematics Education Monograph Number 9* (pp. xx-xx). Reston, VA: NCTM, 98-111.

Clarke, D. J. (Ed.). (2001). *Perspectives on practice and meaning in mathematics and science classrooms*. Dordrecht, Netherlands: Kluwer Academic Press.

Clarke, D. J., Keitel, C., & Shimizu, Y. (Eds.). (2006). *Mathematics classrooms in twelve countries: The insider's perspective*. Rotterdam: Sense Publishers.

Good, T. L. & Brophy, J. E. (2003). *Looking in classrooms*. New York: Allyn & Bacon.

Hattie, J. (2003, October). *Teachers make a difference: What is the research evidence? Distinguishing expert teachers from novice and experienced teacher*. Paper presented at the Building Teacher Quality: What does the research tell us ACER Research Conference, Melbourne, Australia.

Hogan, D., Chan, M., Rahim, R., Kwek, D., Aye, K.M., Loo, S.C., Sheng, Y. Z., & Luo, W. (2013). Assessment and the logic of instructional practice in Secondary 3 English and mathematics classrooms in Singapore. *Review of Education*, 1, 57-106.

Kaur, B. (2009). Characteristics of good mathematics teaching in Singapore Grade 8 classrooms: A juxtaposition of teachers' practice and students' perception. *ZDM Mathematics Education*, 41, 333-347.

Kaur, B. (2017). Mathematics classroom studies: Multiple lenses and perspectives. In Kaiser, G. (Ed.), *Proceedings of the 13th International Congress on Mathematical Education (ICME 13)* (pp 45 – 61). Cham, Switzerland: Springer Open.

Leung, F.K.S. (2006). Mathematics education in East Asia and the West: Does culture matter? In: F. Leung, K-D. Graf & F. Lopez-Real (Eds) *Mathematics education in different cultural traditions: A comparative study of East Asia and the West.* (pp. 21–46). New York: Springer.

Leung, F. K.S. (2001). In search of an East Asian identity in mathematics education. *Educational Studies in Mathematics*, 47(1), 35-41.

Ministry of Education. (2012). *The teaching and learning of 'O' Level, N(A) Level & N(T) Level mathematics.* Singapore, Author.

OECD (2019). *OECD Future of Education and Skills 2030 Concept Note: OECD Learning Compass 2030*, www.oecd.org/education/2030-project/teaching-and-learning/learning/learning-compass-2030/OECD_Learning_Compass_2030_concept_note.pdf.

Perkins, D. (1993). Teaching for understanding. *American Educator*, 17(3), pp. 8, 28–35.

Wiliam, D. (2007). Keeping learning on track: Classroom learning and the regulation of assessment, In: K. Lester (Ed.) *Second handbook of research on mathematics teaching and learning* (1053-1098). Charlotte, NC: Information Age Publishing.

Chapter 5

Mathematical Connections: Beyond Utility

Barry KISSANE

Mathematics is often interpreted as a 'useful' activity, with limited attention paid to its potential to be significant for other reasons. While utility is of course important, it is of diminished significance if students are not engaged with, interested in or attracted to mathematics. Many mathematicians and others over time have drawn attention to the beauty of mathematics and its deep aesthetic qualities, and mathematics is connected richly to our collective cultural heritage. One of the broad aims of mathematics education is to develop positive attitudes towards mathematics; while syllabuses make reference to this broad aim in different ways, it is often hard to see how it is addressed explicitly in official documents. In this chapter, we first consider some ways in which mathematics is connected to a wider world beyond its practical applications. We address in particular the significance of some of the aesthetic aspects of mathematics and its cultural heritage, both its history and contemporary views of its place in the world.

1 Introduction

When mathematical connections are considered, it seems to have become customary in education in recent years for 'useful' connections to be highlighted. There is surely no doubt (at least amongst mathematics teachers) that mathematics is a useful enterprise, and thus significant attention has been paid to the many ways in which it can be applied to practical situations. The 2010 AME Yearbook (Kaur & Dindyal, 2011) was a good example of this, in turn reflecting well the Singapore Mathematics Framework (Ministry of Education, 2012, p.14). The Framework is arranged around mathematical problem solving, and explicitly draws attention to applications and modelling intrinsically involved in problem solving in the real world.

However, it is easy to pay excessive attention to questions of usefulness, presumably in a quest to increase the appeal of mathematics to students. Thus, it is now unusual to see a school mathematics textbook cover that does not highlight applications of mathematics in some way and it seems obligatory for textbook expositions to begin by suggesting that the mathematics offered has practical applications, even when these are a little far-fetched. Nor is it unusual for teachers to begin a fresh topic by alerting students to the utility of mathematics.

These practices are neither surprising nor disturbing, but are perhaps natural responses to the question that is frequently asked by students and others, "When are we ever going to need this?" In this chapter, we will explore instead the place of other sorts of connections than those that focus on utility, in part to provide a balance that reflects the intrinsic nature and appeal of mathematics.

2 *When will we ever use this?*

Over recent years, the question, "When will we ever use this?" seems to have become commonplace in school mathematics classrooms, and seems to have become a preoccupation of many students, teachers, textbook authors and others. It is not clear that the same question is routinely asked of other school disciplines, such as science, art, languages, history and physical education: it seems instead to have become a question distinctively associated with mathematics. As a kind of evidence of this claim, I conducted a Google search of the question (which makes no mention of mathematics at all) and found that all of the resulting links on the first several pages referred to school mathematics.

Of course, one interpretation of this phenomenon might be that students are genuinely interested in the answer, and so are keen to be advised of the utility of mathematics; this would explain some of the prevalence of mathematical applications seen on the covers of textbooks, the availability of posters to make clear the contexts in which mathematics might be found, and the textbook exercises showing 'real-world' contexts. But it remains a mystery why similar questions might not be asked of other school activities.

The question might not be seeking a genuine answer but might instead be a sort of *cris de couer*, a desperate plea for attention from a student drowning in mathematical confusion; it may be easier to distract the teacher with such a question than it is to engage with the mathematics and resolve the problems being faced. The question might be important as a signal to the teacher that a student is disengaged, rather than a genuine enquiry about the ways in which mathematics can be applied to something else.

Alternatively, the question might instead reflect a different need: to understand how mathematics connects with the individuals learning it, assuming that the only possible kind of answer is one concerned with usefulness. Indeed, my suspicion is that thoughtful students also ask about the reasons for learning other subjects in school, but without assuming that the only legitimate responses concern utility.

Jo Boaler (1993) suggested that fresh attention in the UK on contexts for school mathematics emerged in the 1970s, especially for less successful students, helping to build bridges between the mathematics learned in school and the mathematical needs of everyday life. This attention was not restricted to the needs of low attainers, however, and she noted (Boaler, 1993, p.14) that two reasons offered for emphasising the contexts for learning mathematics were as a form of motivation for students and as a device to improve the transfer of learning from mathematics to the real world. Nor was attention of this kind restricted to the UK; for example, Nyabanyaba (1999) discussed similar trends in South Africa while scholars in various other countries noted the trend for the widespread depiction of mathematics in 'real' contexts.

In fact, many of the depictions of mathematics being used in contexts are artificial and illusory, and do not in fact reflect the mathematics that people use in their everyday lives. Thus, while the everyday world often deals with approximations, the world of school mathematics often deals with exact results. Indeed, Boaler (1993) observed:

One difficulty in creating perceptions of reality occurs when students are required to engage partly as though a task were real while simultaneously ignoring factors that would be pertinent in the "real life version" of the task. (p. 14).

In effect, contexts are often used to 'dress up' mathematical questions in quite disingenuous ways. For example, consider the following two questions that were given on a national test in Australia:

A copier prints 1200 leaflets. One-third of the leaflets are on yellow paper and the rest are on blue paper. There are smudges on 5% of the blue leaflets. How many blue leaflets have smudges?

The height of a door is 210 cm. Darren is 5/6 of the height of the door. What is Darren's height?

It is difficult to imagine someone not already knowing how many blue leaflets have smudges in order to determine that there were 5% of them damaged in that way. It is also difficult to imagine that somehow precisely two-thirds of the leaflets were blue or that exactly 5% of them were smudged. Similarly, in what circumstances would someone know that Darren was (exactly!) five sixths of the height of a door, without already knowing his height? Dealing with contexts like this is often a kind of exercise in suspending touch with reality. As Maier (1991, p. 63) noted, such problems are not in fact real, but are merely "school problems, coated with a thin veneer of "real world" associations".

Even when the mathematics is legitimate, many 'uses' described for school mathematics refer to contexts that are of interest mainly to adults, but much less likely to be so to many children. Thus students (who mostly have little money themselves) will learn about the interest earned on term deposits in a bank or even on the financial merits of various housing loan possibilities, many years before the prospect of borrowing money to purchase a home is realistic for them; indeed, for many students some of the 'real-life' contexts in their school mathematics textbook will never in fact be real for them personally, for social and economic reasons.

These concerns are not to suggest that it is improper to draw attention to uses for mathematics; indeed, sometimes it is an excellent and engaging idea. Rather, it seems appropriate to look for other ways in addition to an unrelenting quest for 'usefulness', to help students make mathematical connections.

The Singapore mathematics curriculum framework (Ministry of Education, 2012 p. 14) identifies five attitudes of central importance to student learning: beliefs, interest, appreciation, confidence and perseverance. It seems likely that the emphasis on the usefulness of school mathematics is partly motivated by a concern for the first three of these. In the following sections, we explore briefly the possibility of addressing these three attitudes through other forms of connection.

3 Aesthetics

Many famous mathematicians have attested to a personal aesthetic appeal of mathematics, regarding it as a discipline that provides pleasure in various ways. For example, a widely quoted claim to this effect was made by the English philosopher and mathematician, Bertrand Russell:

> Mathematics, rightly viewed, possesses not only truth, but supreme beauty—a beauty cold and austere, like that of sculpture, without appeal to any part of our weaker nature, without the gorgeous trappings of painting or music, yet sublimely pure, and capable of a stern perfection such as only the greatest art can show.

Similarly, the English pure mathematician G. H. Hardy referred to the aesthetics of mathematics when noting in *A Mathematician's Apology*:

> The mathematician's patterns, like the painter's or the poet's must be beautiful; the ideas, like the colours or the words must fit together in a harmonious way. Beauty is the first test: there is no permanent place in this world for ugly mathematics.

Others have referred to mathematics as a creative endeavor and claimed that its appeal arises from the patterns of various kinds that are involved in the discipline. Indeed, as well as celebrated mathematicians, many teachers have been attracted themselves by the aesthetics of mathematics, so that it seems appropriate that aesthetics might be considered a potentially valuable form of connection for students. It seems essential if students are to be introduced to mathematics that they have experiences that help them to understand its appeal to others (and appreciate that the appeal is not restricted to usefulness for some other purpose).

It is important to recognize that aesthetics is mostly a considerably deeper matter than the simplistic idea that "mathematics is fun", which is sometimes used to sugar-coat mathematical drill and practice activities, using bright colours, cartoon characters and the like. And it is also a more sophisticated concept than merely getting pleasure by being able to perform tasks quickly and efficiently – more than mere successful completion of activities.

Davis and Hersch (1981) observed that aesthetics is of key significance to views about mathematics:

> Blindness to the aesthetic element in mathematics is widespread and can account for a feeling that mathematics is dry as dust, as exciting as a telephone book … contrariwise, appreciation of this element makes the subject live in a wonderful manner and burn as no other creation of the human mind seems to do. (p. x)

While aesthetics has many dimensions, we consider here only two of them, concerned with visual aspects and with intellectual aspects—although frequently these are intertwined in practice.

3.1 *Visual*

One element of aesthetics that is accessible to students, both younger and older, concerns geometric patterns, which have considerable visual appeal. Indeed, such patterns are routinely used in everyday settings such as floor tiles, textiles and interior decoration, although students might not recognize these as mathematical unless attention is drawn to the mathematical features of their designs, including key ideas such as symmetry and tessellation. The decorative feature of mathematics is perhaps most wonderfully expressed in the Islamic world, as a brief visit to almost any mosque or to historically significant buildings such as the Alhambra in southern Spain will attest.

One of the major reasons for students to study tessellations is to appreciate their mathematical basis (the other major reason is to understand the reasoning behind deciding which shapes will tessellate and in which ways.). Of course, students are likely to get a stronger feel for tessellations if they have opportunities to create tessellations for themselves, either through the use of pattern blocks, or paper-and-pencil constructions or even computer versions of these. Figure 1 shows an example of a tessellation with might be made with physical blocks, by drawing or with a computer package, illustrating the surprising fact that any quadrilateral can be used to make a regular tessellation in the plane.

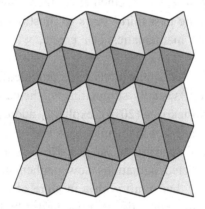

Figure 1: Tessellation with quadrilaterals

Even visual patterns that are too complicated for students to generate personally can give students an aesthetic experience of mathematics. There are many good examples in the book, *Symmetry in Chaos*, generated by Michael Field and Martin Golubitsky (1992), including many richly detailed images that resulted from the use of computer software, together with mathematical algorithms dealing with various aspects of chaos. Field (2019) contains many beautiful examples of this kind, which can be viewed online. The precise details of the mathematics involved in generating the images will be beyond the level of many readers, including school children, but the resulting images are accessible and likely to be appealing to all. Indeed, the authors noted:

> One of our goals for this book is to present the pictures of symmetric chaos—in part because we find them beautiful and in part because we have enjoyed showing them to our friends and think that they may appeal to others. (p. viii)

It seems important for students of mathematics to appreciate that a major publisher would produce a beautiful book to show attractive mathematical patterns, for the sole purpose of sharing beautiful objects with others. In a similar vein, many remarkable images associated with fractals, including the Mandelbrot Set, were included in a very popular book, *The Beauty of Fractals*, (Pietgen & Richter, 1986), one of the first of many books over the past thirty years to illustrate in lavish form the beautiful images that mathematics can create, with the assistance of computer graphics.

In recent years, other publications have provided a rich resource of wonderful examples of visually stimulating mathematics. An especially attractive instance is Pickover (2014), explicitly addressing the aesthetic responses of a wide range of mathematicians and others, together with a stunning collection of appropriate images. He indicates and elaborates through carefully selected quotations how mathematicians through the ages have approached their craft with a mixture of awe, reverence and mystery that verges on the devotional, as suggested by the book's title.

Other elements of the visual and artistic world have also been of interest to mathematics education, including the use of mathematics to understand perspective in painting, the claimed presence of the golden ratio in various works of art, the study of transformations through the medium of Escher drawings, the role of ratios in music, end so on. While such work to help students understand an aesthetic in mathematics is well-intended and generally very interesting, caution is needed, as Sinclair (2001) advised:

> A pernicious consequence of appealing to students' love of something else (whether in the arts, sports, food or money) in the hopes of increasing their interest levels in mathematics is that it endorses the belief that mathematics itself is an aesthetically sterile domain, or at least one whose potentialities are only realized through engagement with external domains of interest. (p. 25)

Accordingly, in the following section, we turn attention to some intrinsically aesthetic aspects of mathematics.

3.2 *Intellectual*

While visual aesthetics are an important element of mathematics, and the most obviously associated with other dimensions of aesthetics, such as painting, design and architecture, and hence the most widely accessible notion of aesthetics, mathematics has an aesthetic appeal in ways that are distinctively different from that of other disciplines. This is probably the meaning of beauty that both Russell and Hardy had in mind with the previous quotes, often referring to the qualities of the reasoning characteristic of mathematics. It is perhaps best captured by the idea of the 'elegance' of a mathematical argument.

A connection between mathematics and beauty is not widely appreciated in modern society, as King (1992) observed:

> Nothing lives further from the intellectual experience of members of the educated public than the notion that mathematics can have aesthetic value. It is remote both to those who are familiar with mathematics and to those who are not. Engineers and scientists, who use mathematics routinely in their work, see it only as a tool. Mathematics, to them, has no more charm than does a microscope or a cloud chamber. Mathematics simply helps them through a days' work. And the humanists, of course, think of mathematics not at all.

Having endured years of required schooling in mathematics, where the subject was presented as something dead as stone and dry as earth and forever separate from their own interests, the humanists have vowed never again to allow it in their presence. (p. 6)

It is perhaps surprising that the aesthetics of mathematics seems to be so little appreciated, especially if, as King (1992, p. 124) claims (and supports with extensive evidence throughout his book) that "Mathematicians do mathematics for aesthetic reasons." Perhaps an extreme example of this claim is the case of the extraordinary mathematician Paul Erdös, whose highest compliment about a colleague's work was that it came 'straight from the book', an imaginary tome that contained all the most elegant and perfect proofs of mathematical results. (Hoffman, 1998, p. 26). Similarly, King reports Lynn Steen's claim:

Despite an objectivity that has no parallel in the worlds of music and art, the motivation and standards of creative mathematics are more like those of art than of science. Aesthetic judgements transcend both logic and applicability in the ranking of mathematical theorems: beauty and elegance have more to do with the value of a mathematical idea than does either strict truth or possible utility. (p. 124)

Reflecting the adage that 'beauty is in the eye of the beholder', opinions of mathematicians about the idea of elegance are often divided. (e.g., Dreyfus & Eisenberg, 1986; Karp, 2008). However, when asked about their favourite examples of the elegance of mathematical thinking, two common choices are Euclid's proof of the irrationality of the square root of two, and that there is an infinite number of prime numbers. Although such work inspired Edna St. Vincent Millay's famous poem, *Euclid Alone has Looked on Beauty Bare*, each of these is a proof by contradiction, and thus an indirect proof and so perhaps less accessible to younger students. But even laypeople can appreciate mathematics from an aesthetic perspective, as reported recently by Johnson and Steinerberger (2019), who found that non-mathematicians used similar criteria for judging mathematical arguments as they used for judging other kinds of artworks, relying mainly on the same dimensions of elegance, profundity and clarity.

Patterns have often been used to develop elegant mathematical arguments, such as that suggested by Figure 2, showing that the sum of successive odd numbers is a square, and used on the cover of Nelsen (1993) as an example of a proof without words (and thus perhaps mixing the visual and the intellectual). Reasoning like this is more likely to appeal to younger students than are indirect proofs.

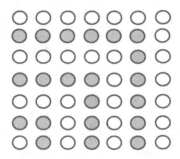

Figure 2: A representation of $1 + 3 + 5 + \ldots + (2n - 1) = n^2$

Similarly the reasoning possible in Figure 3 to show that angles AEC and BED are congruent (since each is the supplement of angle AED) is accessible to young students, just starting geometry, yet giving a sense of a powerful argument used by Euclid about any pair of angles of this kind, not just those shown in the drawing.

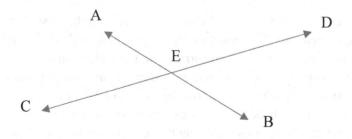

Figure 3: Vertically opposite angles are congruent

Younger children in the primary years can be helped to see the elegance in an argument that the sum of two odd numbers is unavoidably even, rather than relying on considering many separate cases of this phenomenon – and of course, it is never possible in practice to consider all cases. Figure 4 shows a visual representation of the argument, which might also be expressed in other ways. Strong and powerful relationships of this kind might help children see the uniqueness and the elegance of mathematical thought, quite unlike the reasoning encountered in other subjects at school.

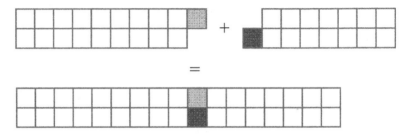

Figure 4: The sum of two odd numbers is an even number

Elegant arguments are not always represented visually, of course. A famous example, accessible to secondary school students is the story of the boy Gauss, finding the sum of the first 100 integers by adding them in a clever way: $(1 + 100) + (2 + 99) + \dots + (50 + 51) = 50 \times 101 = 5050$. Reasoning of this kind is characteristic of the intellectually powerful and satisfying nature of the best mathematical thinking, passing the tests of elegance, profundity and clarity recognised by Johnson and Steinerberger (2019).

However, there is much to be gained by teachers helping students to see and appreciate for the first time the many powerful mathematical relationships in the regular school curriculum, but which can all too often be presented without a suitable sense of surprise and wonder, because they have become familiar already (to the teacher) after many years. For example, while it is difficult for an experienced teacher to recall how surprising it is that the points satisfying the relationship $y = 2x + 1$, lie on a line when plotted on the coordinate plane, it might help students appreciate the power and the wonder of this relationship if we temporarily forget what has become commonplace and see such results from a distant perspective.

As noted by King (1993, p.138), few people recognize mathematics as art, and thus many fail to notice the characteristics of mathematical thinking that give rise to its aesthetic nature. He notes that some of these can be identified in describing the elegance of the Pythagorean proof of the irrationality of root two: "seriousness, depth, generality, unexpectedness, inevitability, and economy" (p. 137). It seems that many students are unlikely to appreciate these qualities of elementary mathematics without systematic and conscientious help from their teacher. It needs to be noted, too, that helping students to appreciate and to enjoy the aesthetics in mathematics is not an easy task, despite its importance. (e.g., Sinclair, 2008).

Separate consideration of aesthetic elements of mathematics (such as the visual and the intellectual) is problematic, as they are frequently intertwined. In a similar way, separation of the aesthetic elements of mathematics from its many cultural and historical roots is also problematic. Indeed, it is at times helpful to consider all of these at once, instead of separately, as reflected in *Quadrivium* (Martineau 2010), arising from the ancient Greek scholars and reflecting their passionate interest in number, named for its four branches of arithmetic, geometry, harmony and astronomy.

We now turn to consider a different element of mathematical connection, concerned with the rich cultural heritage encapsulated by mathematics.

4 Cultural heritage

When considering the development of students' beliefs, interests and appreciation, it seems especially important that the cultural heritage of mathematics be somehow involved. However, in stark contrast to other fields of human endeavor, such as science or the arts, many students appear destined to encounter little of this without deliberate attention of some kind from the teacher and the curriculum. Unlike the worlds of science, entertainment, music, the arts, politics or sport, students do not encounter cultural images of mathematics every day from popular media or their smart phones. Indeed, it may even be worse than that: many students appear to get an impression that mathematics does not have a culture or a history at all, but is rather a soulless space in which humans are neither involved nor welcome: a morass of calculations, formulas, symbols and examinations. Nothing could be further from the truth, however. There are many dimensions to the cultural heritage, past and present, of mathematics, but space here to refer to only selected aspects of some of them.

4.1 *History of mathematics*

Perhaps a natural starting point for considering cultural heritage is to consider the history of mathematics. To do so, of course, is to recognize that mathematics was the product of humans – whether as inventors or as discoverers – and has been a critical part of human civilizations for tens of thousands of years and right up to the present day. It is remarkably easy for students to have no sense of that aspect of our heritage when grappling with computation, number systems, geometry, algebra, trigonometry, calculus, probability, statistics and other mathematical topics included in the curriculum. All of these began through the agency of people, only some of whom are well-known, and an appreciation of mathematics is difficult without some sense of how, when and why these people did what they did.

Attending to the history of mathematics in the school curriculum is not always an easy matter. Although textbooks these days often make brief and passing reference to the historical roots of mathematical ideas and processes, the attention is generally fairly scant and it is rare that a sense of the heritage of mathematical culture is regarded as a proper part of the study of mathematics, or part of its assessment in formal examinations.

Considering the inevitable pressures on time in school curricula, perhaps the best that can be managed in some circumstances is to ensure that students have access to suitable resources and are encouraged to take good advantage of them. Kissane (2009) argues, for example, that popular mathematics materials should be routinely available to students in school libraries as one strategy to do this. Pickover (2009) is an excellent example of such a resource, providing brief information, splendidly illustrated, about many of the most significant mathematical ideas in history and the people responsible for them. Similarly, Strogatz (2019) offers a rich history of the calculus, from ancient times until today, that would provide a much stronger context for students than is possible within a regular classroom. It is hard to imagine a school library that would not be enriched by the inclusion of the remarkable four-volume anthology developed by Newman (1956), even if some of its material is unavoidably a little dated; many of the classic works of the cultural history of mathematics are included within its covers, and mathematics teachers as well as their students will find much of value in it.

Pressures of undergraduate curricula are similar to those of school curricula, too, so that it is not unusual for mathematics teachers themselves to have a limited personal understanding of the history of mathematics, and the sense of perspective that might be provided by such knowledge. As well as published materials, the Internet might provide some assistance to access the history of mathematics in an efficient way (such as to understand the cultural origins of a particular mathematical idea or the significance of a particular person engaged in mathematics). The definitive source for such enquiries is the encyclopedic website maintained regularly by O'Connor & Robertson (2019), which allows for information to be searched in many ways. Less encyclopedic, and less detailed, but possibly more accessible by school students is the website created by Mastin (2010). While the daily classroom might limit the opportunities for the history of mathematics to be dealt with thoroughly, provision of Internet advice for students to explore these in their own time out of school might be one way of assisting them.

4.2 *Diversity and inclusion*

Care is needed in accessing the history of mathematics, in a similar way that care is always needed with accessing written histories, as there is a tendency for the recorded history of mathematics to lack both diversity and inclusion. A common feature involves an emphasis on European contributions, oblivious to those of other cultures. The definitive work countering this view was published by George Gheverghese Joseph (1991), in a direct attempt to counter the 'eurocentrism' he described as characteristic of many published histories and of repeated stories about mathematics. For example, the ancient Babylonians used the Theorem of Pythagoras (although they did not *prove* the theorem) around a thousand years before his birth, and the Persian Omar Khayaam wrote about the binomial coefficients usually referred to as Pascal's Triangle some six hundred years or so before Pascal was born. Similarly, the work of Islamic scholars was generally neglected in conventional histories of mathematics that tend exaggerate the importance of the ancient Greeks and other Europeans, and the role of the Islamic scholars in accessing the earlier work of Chinese and Indian people is often overlooked.

Another notable problem with some versions of the history and the culture of mathematics is the absence of attention to women. In recent years, this neglect has become more prominent, but it is still likely to be the case that students will unwittingly get an impression from textbooks and elsewhere that mathematics is mostly a male enterprise. While there have certainly been systematic prejudices of various kinds that partially explain that impression (such as the refusal to admit women to universities in many countries until relatively recently), care is needed to help students (of both genders) realise that mathematics is not solely a male domain. One way to do this is to bring to students' attention publications such as Parker (1995), which provides strong examples of the life histories and contributions of many women working in mathematics-related fields in the USA. Similarly, the International Organization of Women and Mathematics Education is a useful resource for teachers to become aware of significant developments to support and recognize the work of women whose activities can be most easily accessed via their website (2019).

Regular work has been undertaken for more than thirty years now by the International Study Group on the relations between the History and Pedagogy of Mathematics (often abbreviated to HPM), which conducts conferences, maintains a newsletter and connects a wide range of scholars, teachers and mathematics educators interested in the history of mathematics and in how it might be used productively and practically in education. The HPM website (Tzanakis, 2019) is a rich resource for those wishing to be better informed about recent and forthcoming activities in this area, including three free newsletters annually, often containing interesting articles. A useful publication that grew out of this international collaboration is Katz (2000); while such work is not always directly applicable to the work of teachers in schools, it provides a rich background for teachers, especially for those whose own understanding of the history of their discipline was neglected.

4.3 *The culture of the classroom*

The cultural heritage of mathematics is of course not confined to the past, as mathematics continues to be of key importance to the present world. For most students, the heritage of mathematics is defined in the first instance by the everyday reality of their classroom. While in all countries, mathematics occupies a significant place in the curriculum for all – since mathematics is universally taught to all students, at least for the first ten years of school – there are inevitably differences of opinion about the way in which that is done. In all countries, mathematics curricula are gradually adjusted to accommodate changes in society at large, changes in the mathematical needs of other subjects, technological changes and even personal preferences of senior responsible people; curricula are periodically revised for these and other reasons. Even more significant proposals for change have been presented and discussed, however, and these are likely to be of interest to teachers in connecting their classrooms to mathematics.

For example, Wells (2016) has drawn attention to the neglect of motivation of students in modern curricula (notably in the UK, but his concerns are more wide-ranging). He notes that curricula are frequently overcrowded (as content is added but not removed) and dominated by external examinations (which result in students and teachers focusing attention on success in examinations, instead of on learning, using and enjoying mathematics), with the result that very many students do not enjoy the experience. He suggests that major problems are caused by school curricula trying to do too much too quickly are a lack of attention to proof (the essence of mathematical thinking) and insufficient attention to the ways in which mathematics is used in science, so that students are given limited opportunities to gain an adequate understanding of the nature of mathematics.

Paul Lockhart is a research mathematician in the USA, who recently chose to teach middle years students in schools rather than engage in mathematics research and teaching in universities. His *Mathematician's Lament* (Lockhart, 2009) is also a critique of modern mathematics education, comprising his deeply held opinions that the structure of mathematics in schools is deeply problematic, and set to fail for the great majority of students. Like Wells, he argues that the school mathematics curriculum is imposed on all students, is dominated by excessive structure and examinations, is overloaded with content and results in very few students experiencing the joy of mathematics, or understanding that it is fundamentally a creative activity (and not a rule-following activity) or even understanding what mathematics is actually about. His opinions are strongly counter-cultural, and even emotive, but they offer significant food for thought, as illustrated by the following passage:

> It would be bad enough if the culture were merely ignorant of mathematics, but what is far worse is that people actually think they *do* know what math is about—and are apparently under the gross misconception that math is somehow useful to society! This is already a huge difference between mathematics and the other arts. Mathematics is viewed by the culture as some sort of tool for science and technology. Everyone knows that poetry and music are for pure enjoyment and for uplifting and ennobling the human spirit (hence their almost complete elimination from the public school curriculum), but no, math is *important*. (p. 32)

In a similar vein – although less emotively – Derek Stolp, an experienced mathematics teacher in the USA, has offered deep criticisms of contemporary Western mathematics education, arguing that it is of little real consequence for the great majority of children. (Stolp, 2005).

He further claims that the traditional teaching of mathematics, dominated by externally imposed curricula and assessments does not really help provide tools to understand the world or engage with it, doesn't help students to think logically, doesn't give students a sense of the aesthetics of the discipline or its inherent beauty or promote an awareness of the place of mathematics in the wider culture. (p. 36)

Nor are such criticisms of modern mathematics education particularly new. For example, Richard Skemp, a celebrated British pioneer in the psychology of mathematics education, compared the ways in which mathematics and music are taught, and concluded that major rethinking was necessary:

> If we were to teach children music the way we teach mathematics, we would only succeed in putting most of them off for life. It is by hearing musical notes, melodies, harmonies and rhythms that even the most musical are able to reach the stage of reading and writing music silently in their minds. So why are children still taught mathematics as a paper and pencil exercise which is usually somewhat solitary? ... Music is something which nearly everyone enjoys hearing at a pop, middle-brow or classical level. ... But Mathematicians have only minority audiences, consisting mostly or perhaps entirely of other Mathematicians. The majority have been turned off it in childhood. For these, the music of mathematics will always be altogether silent. (p. 58)

Critiques of these kinds of the implemented mathematics curriculum for very many students in schools deserve serious reflection from teachers and from curriculum developers. While the problems identified are not readily resolved by teachers, there nonetheless seem opportunities for teachers to consider them in framing the daily activity of the classroom. In the final analysis, the most important influence on the culture of a classroom is the teacher, with her perspectives, attitudes, values and practices.

Similarly, the cultural heritage of mathematics would seem to require a sense of how mathematics is related to the world, beyond the relatively unsophisticated and frequently trivial 'applications' of mathematics frequently seen in school mathematics textbooks. One example of this is the field of critical mathematics, concerned with the relationships between mathematics, education and society, and of increasing interest in recent years with widespread unease about the success of conventional mathematics education. Scholars in this field have argued that mathematics teachers should aim their teaching at large issues associated with social change and with social justice, important issues in the lives of many students, rather than with the perpetuation of the status quo. Various key scholars in this emerging field are represented in an excellent collection of essays (Wager & Stinson, 2012).

Finally, a fresh, optimistic and quite different perspective on the place of mathematics in the wider culture was offered recently by Frances Su, in his final address as retiring President of the Mathematical Association of America (Su, 2017):

> So if you asked me: why do mathematics? I would say: mathematics helps people flourish. Mathematics is for human flourishing. ... What I hope to convince you of today is that the practice of mathematics cultivates virtues that help people flourish. These virtues serve you well no matter what profession you choose. And the movement towards virtue happens through basic human desires.

In his address, to be published as a book in 2020, Su elaborates five basic and universal human desires and briefly explores how mathematics is connected with them: play, beauty, truth, justice and love. The Singapore Mathematics model (Ministry of Education 2012) identifies attitudes such as beliefs, interest and appreciation as important for school mathematics, so that reflection on how mathematics in school might attend to these will be supported by a careful reading of this new perspective.

6 Conclusion

While attending to the utility of mathematics will continue to be an important role of both the teacher and the curriculum, it is important to recognise that there are other dimensions of mathematics that also deserve attention in school at an appropriate level. Neglect of these dimensions may undermine other efforts to connect students productively with mathematics. In this chapter, brief sketches of the role of aesthetics and the wider cultural heritage of mathematics have been offered, to support this process.

References

Boaler, J. (1993). The role of contexts in the mathematics classroom: Do they make mathematics more "real"? *For the Learning of Mathematics*, 13(*2*), 12-17.

Davis, P. J. & Hersch, R. (1981) *The Mathematical Experience*. Boston: Birkhaüser.

Dreyfus, T. & Eisenberg, T. (1986) On the aesthetics of mathematical thought. *For the Learning of Mathematics*. 6(*1*), 2-10.

Field, M. (2019) *Images of chaos and symmetry*. Retrieved online 1 August 2019 from https://math.rice.edu/~mjf8/ag/

Field, M. & Golubitsky, M. (1992) *Symmetry in Chaos*. Oxford, UK: Oxford University Press.

Hoffman, P. (1998). *The Man Who Loved Only Numbers*. London: Fourth Estate.

International Organization of Women in Mathematics Education (2019). Retrieved online on 5 August 2019 from https://www.mathunion.org/icmi/organisation/affiliated-organisations/iowme

Johnson, S.G.B & Steinerberger, S. (2019). Intuitions about mathematical beauty: A case study on the aesthetic experience of ideas. *Cognition*, 189, 242-259.

Joseph, G. J. (1991). *The Crest of the Peacock: Non-European Roots of Mathematics*. London: Penguin.

Karp, A. (2008). Which problems do teachers consider beautiful? A comparative study. *For the Learning of Mathematics*. 28(*1*), 36-43.

Katz, V. (2000). *Using history to teach mathematics: An international perspective*. Washington, DC: Mathematical Association of America.

Kaur, B. & Dindyal, J. (2010) (Eds.) *Mathematical applications and modelling: Yearbook 2010 Association of Mathematics Educators*. Singapore: World Scientific.

King, J. P. (1993) *The Art of Mathematics*. New York: Ballantine Books.

Kissane, B. (2009). *Popular mathematics*. In: 22nd Biennial Conference of The Australian Association of Mathematics Teachers, 13-16 July 2009, (pp. 125-134) Fremantle, Western Australia. (Available at https://researchrepository.murdoch.edu.au/id/eprint/6242/)

Lockhart, P. (2009). *A mathematician's lament*. New York: Bellevue Literary Press.

Maier, E. (1991). Folk mathematics. In M. Harris (ed.). *Schools, Mathematics and Work*. (pp 62-66). London: The Falmer Press.

Martineau, J. (ed.) (2010) *Quadrivium: The four classical liberal arts of number, geometry, music and cosmology*. New York: Walker and Company.

Mastin, L. (2010). *The Story of Mathematics*. Retrieved online on 3 August 2019 from https://storyofmathematics.com

Ministry of Education (2012). *Mathematics Syllabus: Secondary 1-4*. Singapore: Ministry of Education.

Nelsen, R. B. (1993) *Proofs Without Words: Exercises in Visual Thinking*. Washington, DC: Mathematical Association of America.

Newman, J. R. (1956) *The World of Mathematics* (four volumes). London: George Allen and Unwin.

Nyabanyaba, T. (1999). Whither relevance? Mathematics teachers' discussion of the use of 'real-life' contexts in school mathematics. *For the Learning of Mathematics*, 19(*3*), 10-14.

O'Connor, J.J. & Robertson, E.F. (2019). *MacTutor History of Mathematics archive*. Retrieved online on 4 August 2019 from https://www-history.mcs.st-andrews.ac.uk/index.html

Parker, M. (Ed.) (1995). *She Does Math! Real-Life Problems from Women on the Job*. Washington, DC: Mathematical Association of America.

Pickover, C. A. (2014) *The Mathematics Devotional*. New York: Stirling.

Pickover, C. A. (2009) *The Math Book*. New York: Stirling.

Pietgen, H.-O. & Richter, P. (1986) *The Beauty of Fractals*. Heidelberg: Springer-Verlag.

Sinclair, N. (2008) Attending to the aesthetic in the mathematics classroom. *For the Learning of Mathematics*. 28(*1*), 29-35.

Sinclair, N. (2001). The aesthetic *is* relevant. *For the Learning of Mathematics*. 21(*1*), 25-32.

Skemp, R. R. (1983) The silent music of mathematics. *Mathematics Teaching*, 102, 58.

Stolp, D. (2005). *Mathematics miseducation: The case against a tired tradition*. USA: Scarecrow Education.

Strogatz, S. (2019) *Infinite Powers*. New York: Houghton Mifflin Harcourt.

Su, F. (2017). Mathematics for human flourishing. Retrieved online on 30 July 2019 from https://mathyawp.wordpress.com/2017/01/08/mathematics-for-human-flourishing/

Tzanakis, C. (2019). *International Study Group on the Relations between the History and Pedagogy of Mathematics (HPM)*. Retrieved online on 12 August 2019 from http://www.clab.edc.uoc.gr/HPM/INDEX.HTM

Wager, A.A. & Stinson, D.W. (2012) (Eds.) *Teaching mathematics for social justice: Conversations with educators.* Reston, VA: National Council of Teachers of Mathematics.

Wells, D. (2016). *Motivating mathematics: Engaging teachers and engaged students.* London: Imperial College Press.

Chapter 6

Using Connecting Mathematical Tasks for Coherence, Connections and Continuity

Leng LOW, Lai Fong WONG

To help students develop a robust understanding of mathematics, the 2020 mathematics syllabus calls for teachers to teach towards big ideas and make visible the *central* ideas, *coherence* and *connections* across topics, and *continuity* across levels. Mathematical tasks are considered to be key to the learning of mathematics as teachers can use mathematical tasks to interact with the students, talk about the mathematics used in the tasks, and make connections among the mathematical ideas within and across levels. Teachers play a key role in selecting mathematical tasks and facilitating productive mathematical discourse that make explicit the mathematical ideas and the connections among them, providing students the experience that learning mathematics is meaningful and worthwhile. In this chapter, the authors will discuss three ways of how a connecting task can be used, with modifications when necessary, for different topics within and across levels to connect various mathematical ideas into a coherent whole.

1 Introduction

The 2020 Singapore mathematics curriculum has called for a shift in presenting mathematics as a coherent and connected body of knowledge, making visible the central ideas and continuity across levels so as to develop in the students a deep and robust understanding of mathematics (Ministry of Education, 2018). This entails teaching towards the identified eight clusters of big ideas in the syllabus that is central to the discipline and helping the students make connections in their mathematical understandings into a coherent whole.

The Singapore mathematics curriculum adopts a spiral approach with an implicit link within and across levels. The spiral approach is to help students make connections across strands over time and develop a deeper understanding of the mathematics as there is continuity across the levels in the mathematical ideas and concepts learned. However, as identified by studies (International Association for the Evaluation of Educational Achievement (IAEEA), 1987; Porter, 1989), a spiral curriculum may be a weakness as "content and goals linger from year to year so that curricula are driven and shaped by still-unmastered mathematics content begun years before" (IAEEA, p. 9). The intent of a spiral curriculum is to add successive depth, but very often, the rapid and superficial coverage of a large number of topics each year leads to fragmented content and compartmentalised understanding of mathematical ideas as the teaching of mathematics topics tends to be taught independently or in isolation from other topics.

Mathematics is a coherent body of knowledge made up of interconnected topics. There are horizontal and vertical connections to be made within and across the levels respectively. The links from one level to the next enable students to have a robust understanding and a better appreciation of mathematics. It is important to examine the learning progressions across the levels to see how the content develops over time (Alberti, 2012). For example, at the primary level, students solve word problems involving numbers and relationships such as ratio and proportion, percentages, rate and speed. The learning of these concepts are further extended at the secondary level, for example, from basic ratio to application of ratios (in secondary one) to map scales as well as direct and inverse proportion (in secondary two). Students consequently deepen their understanding beyond the use of the unitary method to solve problems to the use of algebra and representations of *proportionality* in tables, equations and graphs. Thus, at any one level, teachers must improve focus by tightly linking all the topics for each level, by highlighting the commonalities and differences in the mathematical concepts and ideas learnt at the previous and current levels.

Making connections among different mathematical concepts, their properties and representations form an important part of mathematical understanding. In this chapter, we discuss with exemplars how a connecting task can be used for different topics within and across levels to connect various mathematical ideas into a coherent whole.

2 Theoretical background

2.1 *Making mathematical connections*

Two major types of mathematical connections have been identified in mathematics education literature: (1) recognising and applying mathematics to contexts outside of mathematics (the links between mathematics, other disciplines or the real world), and (2) the interconnections between ideas in mathematics (Blum, Galbraith, Henn & Niss, 2007). In this chapter, we focus on the latter type.

Barmby, Harries, Higgins, and Suggate (2009) suggest that "in order to examine someone's understanding of a mathematical concept, it is important that we examine the connections that a person makes to that concept" (p. 221), and a deep understanding of mathematics is shown through:

- connections made between different mathematical ideas,
- connections made between different representations of mathematical ideas, and
- reasoning between different mathematical ideas.

Students' ability to make connections between apparently separate mathematical ideas is crucial for conceptual understanding (Anthony & Walshaw, 2009b). Connecting mathematical ideas means students are able to link new ideas to related ones and solve challenging mathematical tasks or tasks in new situations by seeking familiar concepts and procedures (Leikin & Levav-Waynberg, 2007).

Researchers (Mhlolo, Schafer & Venkat, 2012; Weinberg, 2001) have argued that interventions might be needed for learners to make connections spontaneously. Ma (1999) advocates that teachers who have understanding of (1) the mathematics in depth, breadth and thoroughness, and (2) the curriculum as a whole rather than in parts that they are required to teach; will then make connections among concepts and procedures, and approach concepts from multiple perspectives in their teaching, demonstrating explicit awareness of "simple but powerful basic concepts" (p. 122). This is also supported by Askew, Brown, Rhodes, Wiliam, and Johnson (1997), whose study reveals that a connectionist approach, with knowledge of the structures and connections within mathematics, contributes to student learning. However, students making mathematical connections cannot be assumed to happen spontaneously – the teacher's intervention is necessary. Teachers thus have to consider the need for learning progression, in terms of task sequencing and demands. If there is no meaningful sequencing of tasks that enables students to come to terms with what they are doing and learning, in the end the demands made on them will not be mathematical – the mathematical task will be reduced to simply completing the task procedurally or copying from the teacher or peers.

2.2 *Learning progression*

Popham (2007) describes a learning progression as a "carefully sequenced set of building blocks that students must master enroute to mastering a more distant curricular aims" and these building blocks "consist of sub-skills and bodies of enabling knowledge" (p. 83). Learning progressions in a subject domain map out student learning outcomes as they advance through different grade levels and a coherent progression of learning "extends previous learning while avoiding repetition and large gaps" (Hunt Institute, 2012, p. 8). Learning progressions guide student learning along "a conceptual corridor in which there are predictable obstacles and landmarks" and "specify at an appropriate and actionable level of detail what ideas students need to know during the development and evolution of a given concept over time" (Confery, 2012, p. 4).

In the learning of mathematics, prior knowledge affects the learning and co-construction of new knowledge. Thus, it is important to establish a proper connection between prior and new knowledge. Understanding the learning progression is crucial for teachers as "it serves as a guiding post for analysing student learning and tailoring their teaching sequence" (Suh & Seshaiyer, 2015). With the understanding of the learning progressions in the syllabuses, in what comes before and after the level, teachers can design mathematical tasks that relate to topics across the different levels, serving as a bridging function for vertical connectivity, so that the depth of students' understanding of the subject matter is not limited to the levels they are in. This chapter focuses on how teachers, through the use of connecting mathematical tasks in a sequential manner, sequence the big blocks in learning and help students make the connections both within and across the levels.

2.3 *Connecting mathematical tasks*

According to Doyle (1983), "tasks form the basic treatment unit in classrooms" (p. 162) because (1) a mathematical task draws students' attention to a particular mathematical concept embedded in it and provides information surrounding that concept; and (2) a mathematical task influences student learning by setting parameters for how the information about the mathematical concept can be processed. Mathematical tasks are central to the learning of mathematics as they can provide a stimulus for students to think about particular mathematical ideas or concepts as well as their connections with other mathematical ideas. Sullivan, Clarke, and Clarke (2013) believe that "learning occurs as a product of students working on tasks, purposefully selected by the teacher" (p. 14). In their research syntheses complemented by evidence from international studies, Anthony and Walshaw (2009b) maintain that effective teachers ensure that "tasks help all students to progress in their cumulative understanding in a particular domain and engage in high-level mathematical thinking" and "posing tasks of an appropriate level of mathematical challenge fosters students' development and use of an increasingly sophisticated range of mathematical thinking and reasoning activities" (p. 155).

In their Task Analysis Guide, Stein, Smith, Henningsen, and Silver (2000) identify two categories of mathematical tasks with high-level cognitive demands, of which one is "procedures with connections tasks". Factors associated with maintenance of high-level cognitive demands include building on students' prior knowledge and making of conceptual connections (Stein, Grover & Henningsen, 1996). According to Anthony and Walshaw (2009a), research findings show that "tasks that require students to make multiple connections within and across topics help them appreciate the interconnectedness of different mathematical ideas and the relationships that exist between mathematics and real life" (p. 15). Literature on teaching and learning of mathematics often recommends solving a mathematical problem or task using different representations or strategies from the same or from different mathematical domains (Levav-Waynberg & Leikin, 2012; Polya, 1963; Schoenfeld, 1985), based on the premise that it develops connectedness of one's mathematical knowledge and deepens mathematics understanding.

Tasks involving multiple representations help students develop both their conceptual understandings and their computational flexibility as they provide students with opportunities to translate between the array of representations. Anthony and Walshaw (2009a) give an example that "a student working with different representations of functions (real-life scenarios, graphs, tables, and equations) has different ways of looking at and thinking about relationships between variables" (p. 16). Solving a task in different ways help students develop connectedness in mathematical ideas. Stigler and Hiebert (1999) find that reinforcing the idea that there may be multiple solutions to a problem or task enhances the quality of students' learning. When solving a task in different ways, the construction of mathematical knowledge is supported by shifting between various representations and connecting different mathematical concepts and ideas (Boaler, 1998; Schoenfeld, 1988; Silver, Ghousseini, Gosen, Charalambous & Strawhun, 2005). Essentially, solving a mathematical problem or task in multiple ways helps students construct mathematical connections as "different solutions can facilitate connection of a problem at hand to different elements of knowledge with which a student may be familiar, thereby strengthening networks of related ideas" (Silver et al., 2005, p. 228).

Leikin and Levav-Waynberg (2007) define a *multiple-solution connecting task* (CT) as "one that may be attributed to different topics or to different concepts within a topic of the mathematics curriculum, and therefore may be solved in different ways" (p. 350). In their work, they consider CTs based on three types of mathematical connections in their work: (a) Connections based on similarities and differences between various representations of the same concept; (b) Connections between different mathematical concepts and procedures and (c) Connections between different branches of mathematics. Drawing upon their work, we are suggesting a particular type of mathematical tasks, which we will also call *connecting tasks*, that (1) require students to make connections to and among mathematical concepts and representations, and (2) are used across topics within or across levels.

3 Examples of connecting tasks

In this section, we describe three examples of *connecting tasks* that teachers can used within or across levels to allow students to make the following connections:

1. Connections between various representations of the same mathematical concept
2. Connections between different mathematical concepts and procedures
3. Connections between different strands of mathematics

3.1 *Task connecting between various representations of the same mathematical concept*

Folding a box from cardboard (see Figure 1) is an example of a task that connects between various representations of the same mathematical concept across different levels.

A sheet of cardboard measuring 30 cm by 20 cm is to be cut and folded into an open rectangular box. A square from each corner of the cardboard, (shown as shaded parts) will be cut out and the remaining portion of the cardboard will be folded to make an open box.

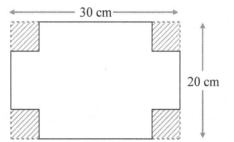

By changing the size of the square cut out, the volume of the box can be changed. What is the size of square to be cut so as to maximise the volume that the box could hold?

Figure 1. Folding a box from cardboard.

At secondary 1, students can approach this task using the listing heuristic that they have learnt in primary school. From the table of values (Figure 2), students may conclude that cutting out squares of size 4cm will maximise the volume that the box could hold. However, the size of the square considered thus far has been an integer, and so the teacher has to prompt students to explore further values, say, to 1 decimal place, that will maximise the volume of the box. At this juncture, it may be an opportune moment to introduce the idea of computational thinking where students can leverage the power of Excel spreadsheet to tabulate the necessary values. This set of values of the size of square cut and the volume of box can be preserved for further exploration at subsequent levels.

Size of square cut (cm)	Length of box (cm)	Width of box (cm)	Height of box (cm)	Volume of box (cm^3)
1	28	18	1	504
2	26	16	2	832
3	24	14	3	1008
4	22	12	4	1056
5	20	10	5	1000
6	18	8	6	864
7	16	6	7	672
8	14	4	8	448
9	12	2	9	216

Figure 2. Table of values obtained using the listing heuristic.

At secondary 2, students can revisit this task when they learn about functions and graphs – the concept of (and big ideas about) "functions as a relationship between two variables that expresses how one variable can be determined from another and how a change in one can affect the other" (MOE, 2018, p. 3A-4). Students have learnt about linear graphs in secondary 1, and are able to extend their knowledge of graphs to non-linear ones. They can use the set of values obtained in secondary 1 to plot a graph and use it to solve the task. Through the task, students make sense of how graphs are used to represent the relationships between two sets of values, and appreciate how graphs help them visualise the relationships between the variables, before they next connect it to some algebraic equation. We suggest this task be used as a trigger to introduce the idea of non-linear functions before they formally learn quadratic function. From this activity, students learn that a function can be represented in multiple equivalent forms, for example, in tables, or in the form of diagrams such as graphs; and that functions can be used as models of many real-world situations and phenomena. Opportunities should be provided for students to compare and discuss the two approaches they have used at the two levels.

At upper secondary, the students can approach the same task using calculus. The table of values and the graph obtained in the previous levels serve to help them make sense of, and make connections between, the various representations and mathematical ideas. This optimisation task serves as a good example of how they apply the derivatives to solve problems in real-world contexts. An orchestrated discussion on the various approaches and representations used at various levels is crucial to help students make the connections.

3.2 *Task connecting between different mathematical concepts and procedures*

Finding the height of the flag pole, a task shared by Low and Wong (2019), connects various different mathematical concepts and procedures across different levels. Figures 3a, 3b, and 3c show how the task to find the height of the flag pole can be posed to students at secondary 1, 2 and 3 respectively.

Instructions:
1. Write down your height.
2. Now, take a picture of you standing next to the flag pole.
3. Describe and explain how you can use this picture to get an estimate of the height of the flag pole. You may want to draw a suitable diagram to make your explanation clearer. State any assumption(s) made.

Figure 3a. Finding the height of the flag pole at secondary 1.

Instructions:
1. Write down your eye's height.
2. Now, place a mirror on the ground at __ metres from the foot of the flag pole. Looking at the mirror, walk backwards, in line with the flag pole and the mirror, till you see the tip of the flag pole in the mirror. Measure the horizontal distance between you and the mirror.
3. Describe how you can use this distance to get an estimate of the height of the flag pole. You may want to draw a suitable diagram to make your explanation clearer. State any assumption(s) made.

Figure 3b. Finding the height of the flag pole at secondary 2.

Instructions:
1. Write down your eye's height.
2. Now, download a clinometer app on your smartphone.
3. Describe how you can use the clinometer to get an estimate of the height of the flag pole. You may want to draw a suitable diagram to make your explanation clearer. State any assumption(s) made.

Figure 3c. Finding the height of the flag pole at secondary 3.

When working on the task at secondary 1, students activate their proportional thinking to set up an equation of two equivalent ratios with the height of the flag pole as a missing term. The idea of proportionality and equivalence can be weaved in a discussion, orchestrated by the teacher, on the two possible equations set up using different pairs of ratios:

$$\frac{\text{Actual height of flag pole}}{\text{Actual height of person}} = \frac{\text{Height of flag pole in picutre}}{\text{Height of person in picture}}$$

and

$$\frac{\text{Actual height of flag pole}}{\text{Height of flag pole in picture}} = \frac{\text{Actual height of person}}{\text{Height of person in picture}}$$

The students are then motivated to look forward to the learning of the concept of scales in the following level and see how the mathematical ideas of ratios and scales are connected.

When working on the same task at secondary 2, the students use a different mathematical approach – one that involves similar triangles. An orchestrated discussion on the approaches used in the current and previous levels further helps them make connections between the mathematical concepts, skills and ideas used in the two approaches. They may then be motivated to look forward to the learning of similarity tests in the following level when their intuitive notions about similarity can be proven.

At secondary 3, the students can revisit the similar triangles identified during the task in the previous level and use their current knowledge of similarity tests to prove the similar triangles identified previously. An orchestrated discussion on the various approaches used in different levels helps students make connections between the various mathematical concepts, skills and ideas involved, via the big idea about *proportionality*.

3.3 Task connecting between different strands of mathematics

Proving Pythagoras' Theorem is an example of a task that connects between different strands of mathematics across different levels, namely, the Numbers and Algebra strand and the Geometry strand.

At secondary 1, before learning about Pythagoras' Theorem and its proofs, students explore the following relationship:

> *The difference of the squares of any two consecutive numbers is always an odd number.*

They will discover a special set of odd numbers, which are perfect squares that is a difference of two squares, in other words, $a^2 = c^2 - b^2$. This leads to a special set of positive integers a, b and c, known as the Pythagorean Triple, that fits the rule $a^2 + b^2 = c^2$. Students can continue to explore further facts about the Pythagorean Triple:

- A Pythagorean Triple always consists of (i) all even numbers, or (ii) two odd numbers and an even number.
- A Pythagorean Triple can never be made up of all odd numbers or two even numbers and one odd number. It is easy to construct sets of Pythagorean Triples.
- Every odd number, except 1, is the smallest of a Pythagorean Triple.
- a, b and c form a Pythagorean Triple if $a = m^2 - n^2$, $b = 2mn$, and $c = m^2 + n^2$, where m and n are any two positive integers ($m > n$).

At secondary 2, they revisit the above facts and prove them using the following algebraic identities they have learnt:

- $(a \pm b)^2 = a^2 \pm 2ab + b^2$
- $(a + b)(a - b) = a^2 - b^2$

Subsequently, when they learn the Pythagoras' Theorem, they discover that a Pythagorean Triple forms the sides of a right-angled triangle. Students can continue to investigate other non-integer values which satisfy the relationship $a^2 + b^2 = c^2$ and thus form the sides of a right-angled triangle. At this level, they are encouraged to take a more formal approach towards geometry, and to some degree of deductive reasoning, prove the theorem as part of their learning experiences. An example of a proof is shown in Figure 4a.

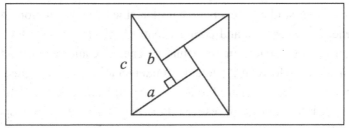

Figure 4a. An approach to prove Pythagoras' Theorem in secondary 2.

The students have to communicate their geometrical results – (1) why placing four congruent right-angled triangles in this way forms a big square and a small square, and (2) how the relationship $a^2 + b^2 = c^2$ is obtained – in the form of logical and concise explanations and arguments, using appropriate geometrical properties and algebraic identities. [See Annex A for this proof.]

After learning the idea of similarity in secondary 2, they can revisit the Pythagoras' Theorem to prove it again, by first identifying and proving the similar triangles (Figure 4b), and then using the properties of similar triangles to derive the relationship $a^2 + b^2 = c^2$. [See Annex B for this proof.]

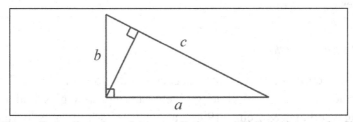

Figure 4b. Another approach to prove Pythagoras' Theorem in secondary 3.

As proposed by Low and Wong (2019), "learning is neither uniform nor uni-directional" (p. 377). Using a connecting task that requires students to make connections to and among mathematical concepts and representations, within/across topics and within/across levels, enables students to reinforce, construct and consolidate their learning by looking back at what they have learnt or understood, looking at what they are learning, and looking forward to what they will be learning. Learning advances when students revert to familiar objects and situations as new experience emerges, use and connect mathematical ideas, strategies and techniques across topics, and eventually generalise and reconstruct ideas for themselves. However, students' understanding and progression in mathematics are not measured in terms of completed tasks – a datum that teachers must be mindful of, especially if the students merely respond to tasks without contacting the mathematical thinking and connecting the mathematical ideas as intended. The sequence of tasks may not develop the students' understanding of the mathematical ideas if the teacher does not make meaningful connections between the topics across the levels through purposeful questioning and facilitation.

4 Some issues to address

Using connecting tasks for coherence and continuity in the mathematics classrooms no doubt places new demands on teachers – the selection of a suitable task and the maintenance of the rigour of the task during instruction. Increasing students' exposure to and sustained engagement in such tasks requires changes in the knowledge and instructional practices of mathematics teachers.

4.1 *Selection of tasks*

In the selection of tasks, the teacher needs to consider tasks that encourage the use of multiple problem solving strategies and extend thinking (Spillane & Zeuli, 1999), as well as tasks that help students make connections among the various mathematical representations (Kaput, 1989). The mathematical-task knowledge required of the teacher includes: (1) the understanding of the nature of connecting tasks, such as: the mathematics content; multiple approaches to solve the problem;

connection to other mathematical ideas; requiring students to justify and conjecture; high cognitive demand (Stein et al., 2000); (2) the ability to identify, select and/or create connecting tasks that afford the learning of mathematics with deep understanding (Chapman, 2013); (3) the knowledge of "what aspects of a task to highlight, how to organize and orchestrate the work of the students, what questions to ask to challenge those with varied levels of expertise and how to support students without taking over the process of thinking for them and thus eliminating the challenge" (NCTM, 2000, p.19).

Connecting tasks require much from both the teacher and the students. The learning process involves more effort and deeper thinking in the teacher and students, as simply following a set of steps is not an approach that will lead to success or to understanding. To engage students in thinking mathematically and making mathematical connections, the connecting task used must also bring forth the learning experiences that:

- include multiple entry points for students to access the mathematics;
- provide challenge for all students to extend their thinking;
- allow students to actively make sense of mathematics through mathematical reasoning; and
- allow students opportunities to express their understanding in different ways or reach an understanding via different strategies.

(Anthony & Walshaw, 2009a)

Connecting tasks can vary not only with respect to mathematics content but also with respect to the cognitive processes involved in working on them (Stein et al., 1996). These tasks offer students the opportunity to extend what they know and stimulate their learning. Besides considering the nature of the tasks, attention to the classroom processes associated with mathematical tasks (such as making conjectures, reasoning, validating the assertions, and discussing and questioning own thinking and the thinking of others) is equally needed. Hence, it entails a shift in teachers' effort to (1) attend to students' thinking; (2) make public their efforts to elicit student thinking; and (3) recognise students' mathematical competencies (Kazemi & Franke, 2004).

4.2 *Enactment of tasks*

Besides the selection of connecting tasks, how a task is enacted to bring about connections is equally important. During the enactment of a task, the role of the students is to find approaches to solve it, identify the mathematical concepts and skills, and make the connection between the mathematical ideas. The role of the teacher is to monitor, facilitate and orchestrate whenever appropriate to ensure that the intended student learning happens.

Research has shown that it is crucial for the teacher to orchestrate the implementation of a rich task to ensure students are using the task to promote understanding through mathematical thinking, reasoning, and problem-solving (e.g. Stein & Lane, 1996). Stein and Kaufman (2010) argue that mathematical tasks create the potential for students' learning but whether or not the tasks achieve their potential depends very much on how teachers enact them. The teachers have to pay attention to what students do and say as they work on the task so as to be able to understand their mathematical thinking and make them visible.

When students are actively engaged in the task, teachers must take the opportunity to effectively monitor students' learning – notice students' thinking, provide opportunities for rich questioning, and prepare for subsequent discussion and lesson consolidation. More importantly, the teacher has to identify the mathematical learning potential of particular strategies or representations used by the students, and carefully select and sequence student responses to share with the class as a whole during the discussion phase (Stein, Engle, Smith, & Hughes, 2008). As students and the teacher consider, question and add to each other's thinking, important mathematical ideas and connections are co-produced. Hearing students' ideas can also give teachers a window into their thinking, and also the conceptions and misconceptions underlying students' understanding. The final stage of discussion and consolidation plays a significant role in 'explicitising' the mathematical connections, deepening learning and building confidence in students.

To effectively enact the task, the teacher has to think about the mathematics involved and the mathematical thinking needed in the task. This requires specific teacher knowledge to facilitate the eliciting of the mathematics during students' discussion. According to Ball (1990), teachers need to acquire "the ability to think with precision about mathematical tasks and their use in class" so that they can "select, modify and enact mathematical tasks with their students" (Stein et al., 2000, p. xii).

4.3 *Teacher's knowledge*

In order to help students make different possible mathematical connections using connecting tasks, teachers play a crucial role in this intervention as they must facilitate the learning process in ways that will enable learners to recognise and make sense of these mathematical connections. Classroom instruction organised by teachers depends very much on teachers' knowledge and beliefs about mathematics as well as on their understanding about mathematics teaching and learning (Anthony & Walshaw 2009b).

Having good content knowledge allows teachers to represent mathematics as a coherent and connected system (Ball & Bass, 2000). Teachers with good content knowledge are also able to assess their students' current level of understanding and use their knowledge to make decisions regarding mathematical tasks, classroom discourse and resources to feed forward into the learning process. In order to help students construct meaningful knowledge structures, teachers themselves must possess richly connected understandings and content knowledge in their subject matter (Ball, 1990).

Teachers' classroom practices are usually derived from their understanding of the curriculum and their role as mathematics teachers in this learning process. As students' understanding may differ from those planned by the teachers, it is important for the teachers to listen in to the students' solutions and be flexible in their interactions with them.

Sometimes, the teachers may learn from and learn with the students. Thus, teachers need to have the belief that learning is an ongoing process and our teacher knowledge in terms of content knowledge, pedagogical knowledge and knowledge of the learners has to grow continuously.

5 Conclusion

In this chapter, we look at how connecting tasks can be used to present students with a logically sequenced and connected curriculum that builds on skills and concepts learnt before. Looking at the demands placed on the teachers when planning for and enacting such tasks in the classroom, professional development initiatives has to be provided in the areas of:

- the selection of mathematical tasks,
- the development of skills needed to enact the tasks, and
- the enhancement of the teachers' knowledge.

The teacher plays an important role in the selection of a connecting task. It involves a decision-making process which can be complex and teacher development should focus on helping teachers understand this complexity (Sullivan & Mousley, 2001). Teachers' pedagogical content knowledge, as defined by Shulman (1986), is enhanced when the professional development efforts supported them in examining the complexities of decision-making with regard to connecting tasks. While there are well-planned choice of and learning goals of the task, teachers need to hone their skills in working out how they can best help students grasp and connect mathematical ideas (Hill, Rowan, & Ball, 2005). Professional development for teachers should focus on developing teachers to make on-the-spot classroom decision making so that they can perform more finely tuned listening and questioning so as to leverage and build on students' responses to extend their learning. Silver et al. (2005) opine that limitations in the teachers' mathematical knowledge may hamper the use of mathematical tasks to make connections in this teaching and learning process. An effective teachers' professional development should include content as the focus, sustained over time, collective participation and collaboration of teachers working on issues central to students' learning, organized around the instructional materials to be used in the classroom (Ball, Sleep, Boerst, & Bass, 2009).

Kazemi (2008) argued that improvements in students' learning outcomes will require support and resourcing which can come from the joint efforts of everyone involved in mathematics education – none other than the mathematics teachers within the school. A suggestion to effective implementation of connecting tasks is the use of vertical teaming, as advocated by Suh and Seshaiyer (2015), where a team is formed by teachers teaching across the levels. In this arrangement, each level teacher has a crucial role to play in the student's mathematical development, as every teacher needs to think about the level and also beyond the level taught. Vertical teaming and a focus on the learning progressions through the use of connecting tasks, provide opportunities for teachers to discuss the expectations of the mathematics required and the importance of breaking down the essential mathematical learning across the levels.

References

Alberti, S. (2012). Making the shifts. *Educational Leadership, 70*(4), 24-27.

Anthony, G., & Walshaw, M. (2009a). *Effective Pedagogy in Mathematics* (Vol. 19). Belley, France: International Academy of Education.

Anthony, G., & Walshaw, M. (2009b). Characteristics of Effective Teaching of Mathematics: A view from the West. *Journal of Mathematics Education, 2*(2), 147-164.

Askew, M., Brown, M., Rhodes, V., Wiliam, D., & Johnson, D. (1997, September). *Effective teachers of numeracy in primary schools: Teachers' beliefs, practices and pupils' learning.* Paper presented at the British Educational Research Association Annual Conference, University of York.

Ball. D. L. (1990). Prospective elementary and secondary teachers' understanding of division. *Journal for Research in Mathematics Education, 21*(2), 132-144.

Ball, D. L., & Bass, H. (2000). Interweaving content and pedagogy in teaching and learning to teach: Knowing and using mathematics. In J. Boaler (Ed.), *Multiple perspectives on mathematics of teaching and learning*. (pp. 83-104). Westport, Conn.: Ablex Publishing.

Ball, D. B., Sleep, L., Boerst, T. A., & Bass, H. (2009). Combining the development of practice and the practice of development in teacher education. *The Elementary School Journal, 109*(5), 458-474.

Barmby, P., Harries, T., Higgins, S., & Suggate, J. (2009). The array representation and primary children's understanding and reasoning in multiplication. *Educational Studies in Mathematics, 70*(3), 217-241.

Blum, W., Galbraith, P. L., Henn, H. W., & Niss, M. (2007). *Modelling and applications in mathematics education: the 14th ICMI study*. New York: Springer.

Boaler, J. (1998). Open and closed mathematics: Student experiences and understandings. *Journal for Research in Mathematics Education, 29*(1), 41-62.

Chapman, O. (2013). Mathematical-task knowledge for teaching. *Journal of Mathematics Teacher Education, 16*(1), 1-6.

Confery, J. (2012). Articulating a learning science foundation for learning trajectories in the CCSS-M. In L. R. Van Zoerst, J. J. Lo, & J. L. Kratky (Eds.), *Proceedings of the 34th annual meeting of the North American Chapter of the International Group for the Psychology Mathematics Education* (pp. 2-20). Kalamazoo, MI: Western Michigan University.

Doyle, W. (1983). Academic work. *Review of Educational Research, 53*(2), 159-199.

Hill, H. C., Rowan, B., & Ball, D. L. (2005). Effects of teachers' mathematical knowledge for teaching on student achievement. *American Educational Research Journal, 42*(2), 371-406.

Hunt Institute. (2012). Advance America: A commitment to education and the economy. Issue Briefs. Prepared for the 2012 governors education symposium. Retrieved from http://www.hunt-institute.org/knowledge-library/articles/2012-5-24/2012-governors-education-symposium-briefs/.

International Association for the Evaluation of Educational Achievement. (1987). *The underachieving of curriculum: Assessing U.S. school mathematics from an international perspective*. Urbana: University of Illinois.

Kaput, J. J. (1989). Linking representations in the symbol systems of algebra. In S. Wagner, & C. Keiran (Eds.), *Research issues in the learning and teaching of algebra* (pp. 167-194). Hillsdale, NJ: Lawrence Erlbaum Associates.

Kazemi, E. (2008). School development as a means of improving mathematics teaching and learning: Towards multidirectional analyses of learning across contexts. In K. Krainer, & T. Wood (Eds.) *The Handbook of Mathematics Teacher Education: Volume 3* (pp. 207-230). Brill Sense.

Kazemi, E., & Franke, M. L. (2004). Teacher learning in mathematics: Using student work to promote collective inquiry. *Journal of Mathematics Teacher Education, 7*(3), 203-235.

Leikin, R., & Levav-Waynberg, A. (2007). Exploring mathematics teacher knowledge to explain the gap between theory-based recommendations and school practice in the use of connecting tasks. *Educational Studies in Mathematics, 66*(3), 349-371.

Levav-Waynberg, A., & Leikin, R. (2012). The role of multiple solution tasks in developing knowledge and creativity in geometry. *The Journal of Mathematical Behavior, 31*(1), 73-90.

Low, L., & Wong, L. F. (2019). Making vertical connections when teaching towards big ideas. In T. L. Toh, & B. W. J. Yeo (Eds.). *Big ideas in mathematics: Yearbook 2019* (pp. 367-383). Association of Mathematics Educators.

Ma, L. (1999). *Knowing and teaching elementary mathematics: Teacher's understanding of fundamental mathematics in China and the United States.* Lawrence Erlbaum Associates, Incorporated.

Ministry of Education. (2018). *2020 secondary mathematics syllabus (draft).* Singapore: Curriculum Planning and Development Division.

Mhlolo, M. K., Schafer, M., & Venkat, H. (2012). The nature and quality of the mathematical connections teachers make. *Pythagoras, 33*(1), 1-9.

National Council of Teachers of Mathematics. (2000). *Principles and standards for school mathematics* (Vol. 1). National Council of Teachers of Mathematics.

Polya, G. (1963). On learning, teaching, and learning teaching. *The American Mathematical Monthly, 70*(6), 605-619.

Popham, W. J. (2007). The lowdown on learning progressions. *Educational Leadership, 64*(7), 83-84.

Porter, A. (1989). A curriculum out of balance: The case of elementary school mathematics. *Educational Researcher, 18*(5), 9-15.

Schoenfeld, A. H. (1985). *Mathematical problem solving.* New York: Academic Press.

Schoenfeld, A. H. (1988). When good teaching leads to bad results: The disasters of 'well-taught' mathematics courses. *Educational Psychologist, 23*(2), 145-166.

Shulman, L. S. (1986). Those who understand: Knowledge growth in teaching. *Educational Researcher, 15*(2), 4-14.

Silver, E. A., Ghousseini, H., Gosen, D., Charalambous, C., & Strawhun, B. T. F. (2005). Moving from rhetoric to praxis: Issues faced by teachers in having students consider multiple solutions for problems in the mathematics classroom. *Journal of Mathematical Behavior, 24*(3-4), 287-301.

Spillane, J. P., & Zeuli, J. S. (1999). Reform and teaching: Exploring patterns of practice in the context of national and state mathematics reforms. *Educational Evaluation and Policy Analysis, 21*(1), 1-27.

Stein, M. K., Engle, R. A., Smith, M. S., & Hughes, E. K. (2008). Orchestrating productive mathematical discussions: Five practices for helping teachers move beyond show and tell. *Mathematical Thinking and Learning, 10*(4), 313-340.

Stein, M. K., & Kaufman, J. H. (2010). Selecting and supporting the use of mathematics curricula at scale. *American Educational Research Journal, 47*(3), 663-693.

Stein, M. K., Grover, B., & Henningsen, M. (1996). Building student capacity for mathematical thinking and reasoning: An analysis of mathematical tasks used in reform classrooms. *American Educational Research Journal, 33*(2), 455-488.

Stein, M. K., & Lane, S. (1996). Instructional tasks and the development of student capacity to think and reason: An analysis of the relationship between teaching and learning in a reform mathematics project. *Educational Research and Evaluation, 2*(1), 50-80.

Stein, M. K., Smith, M. S., Henningsen, M. A., & Silver, E. S. (2000). *Implementing standards-based mathematics instruction: A casebook for professional development.* New York: Teachers College Press.

Stigler, J. W., & Hiebert, J. (1999). *The teaching gap: Best ideas from the world's teachers for improving education in the classroom.* New York, NY: The Free Press.

Suh, J., & Seshaiyer, P. (2015). Examining teachers' understanding of the mathematical learning progression through vertical articulation during lesson study. *Journal of Mathematics Teacher Education, 18*(3), 207-229.

Sullivan, P., Clarke, D., & Clarke, B. (2013). *Teaching with tasks for effective mathematics learning* (Vol. 9). Springer Science & Business Media.

Sullivan, P. A., & Mousley, J. (2001). Thinking teaching: Seeing mathematics teachers as active decision makers. In F-L. Lin & T. J. Cooney (Eds.), *Making sense of mathematics teacher education* (pp. 147-163). Springer, Dordrecht.

Weinberg, S.L. (2001, April). *Is there a connection between fractions and division? Students' inconsistent responses.* Paper presented at the Annual Meeting of the American Educational Research Association, Seattle, WA.

Annex A

An approach to prove Pythagoras Theorem in secondary 2.

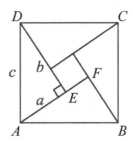

First, to prove that placing the four congruent right-angled triangles in this way forms a big square ABCD.

Consider triangles *ADE* and *ABF*.

Let $\angle DAE = \theta$.

Then $\angle ADE = 90° - \theta$ ($\angle$ sum of Δ)

Since $\Delta ADE \equiv \Delta ABF$, $\angle BAF = \angle DAE = \theta$

         and $\angle ABF = \angle ADE = 90° - \theta$

Thus, $\angle DAB = \theta + (90° - \theta)$

         $= 90°$

Similarly, $\angle ABC = \angle BCD = \angle CDA = 90°$

Hence, *ABCD* is a square.

Next, to prove the relationship $a^2 + b^2 = c^2$.

Area of big square $=$ area of 4 right triangles + area of small square

$$c^2 = 4 \times \frac{1}{2}ab + (b-a)^2$$
$$= 2ab + (b^2 - 2ab + a^2)$$
$$= a^2 + b^2$$

Annex B

Another approach to prove Pythagoras Theorem in secondary 3.

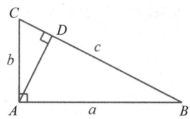

In right-angled triangle ABC, construct a perpendicular from A to BC.

First, to prove that the three triangles are similar.
Consider triangle ABC.
Let $\angle ABC = \theta$.
Then $\angle ACB = 90° - \theta$ ($\angle$ sum of Δ)
Consider triangle DBA.
     $\angle DBA = \angle ABC = \theta$ (common)
and $\angle DAB = 90° - \theta$ since $\angle BAC = 90°$
Consider triangle DAC.
     $\angle DCA = \angle ACB = 90° - \theta$ (common)
and $\angle DAC = 90° - (90° - \theta)$ ($\angle$ sum of Δ)
       $= \theta$
Thus, $\angle ABC = \angle DAC = \angle DAB = \theta$
and $\angle ACB = \angle DAC = \angle DAB = 90° - \theta$
By AA Similarity Test, triangles ABC, DBA and DAC are similar.

Next, to prove the relationship $a^2 + b^2 = c^2$.

Since triangles ABC and DBA are similar,

$$\frac{AB}{DB} = \frac{BC}{BA}$$

$\Rightarrow \quad AB^2 = BC \times DB \qquad \ldots\ldots\ldots (1)$

Since triangles ABC and DAC are similar,

$$\frac{AC}{DC} = \frac{BC}{AC}$$

$\Rightarrow \quad AC^2 = BC \times DC \qquad \ldots\ldots\ldots (2)$

Adding equations (1) and (2),

$$AB^2 + AC^2 = BC \times DB + BC \times DC$$
$$= BC \times (DB + DC)$$
$$= BC \times BC$$
$$= BC^2$$

i.e. $\quad a^2 + b^2 = c^2$

Chapter 7

Making Connections: Teaching through Problem Solving

Joseph Kai Kow YEO

The Singapore primary school mathematics curriculum defines connection as the ability to see and link mathematical ideas, between mathematics and other subjects, and between mathematics and the real world. Teachers can facilitate their students to make mathematical connections by teaching through problem solving. Teaching through problem solving involves teaching new mathematics concepts with problem solving contexts and enquiry-oriented environments. In this chapter, the notion of teaching through problem solving is discussed and illustrated through developing a lesson teaching structure that connects with problem-solving strategies. A sample lesson plan on teaching through problem solving for a primary 4 class on area and perimeter of a rectangle is illustrated.

1 Introduction

Mathematical Problem Solving has been the central focus of Singapore mathematics curriculum for more than twenty years. In order to solve various type of mathematical problems, particularly, the process problems, a pupil has to apply four types of mathematical abilities, namely, specific mathematical concepts, skills, processes, and metacognition. The most recent primary mathematics syllabus, which was released in 2019 and will be implemented in 2021, continues to retain mathematical problem solving as the central focus and list one of the general aims for mathematics education as to "develop thinking, reasoning, communication, application and metacognitive skills through a mathematical approach to problem solving" (Ministry of Education, 2019 p. 8). It also explicitly encourages teachers to create opportunities for pupils to justify their answers, both classroom discussion and in written work (Ministry of Education, 2019).

However, most mathematical instructions in Singapore consist of the traditional whole class expository teaching followed by pupil practice of routine exercises, and regular written tests comprising multiple-choice questions, short-answer and long-answer open response questions (Chang, Kaur, Koay, & Lee, 2001). While low-level skills are easy to show and assess, excessive use of low-level skills misses valuable learning opportunities such as contributing and challenging mathematical ideas (Goos, 2004). Yeo and Zhu (2005), who used the data from 118 coded mathematics lessons from 18 Primary five classes and 19 Secondary three classes in Singapore, revealed that mathematics teaching in these classes was mainly teacher-centred with a focus on memorisation and acquisition of basic concepts and routine procedural skills. They concluded that there was much room for the integration of higher-order thinking skills in Singapore mathematics classrooms. The implication of this transformation in perspective is to support pupils to become effective problem solvers. Mathematics teachers need to change their interpretation of problem solving as an additional topic in the mathematics curriculum. Instead, pupils must have the opportunity to learn mathematics through problem solving (Hiebert, Carpenter, et al. 1997; Hiebert, Gallimore, et al. 2003; Hiebert & Wearne 2003; Mason, Burton, and Stacey 1985; NCTM 2000; Pólya 1945; Schoen and Charles, 2003; Schoenfeld, 1985). This approach embraces the understanding that there is a synergetic connection between problem solving and concept learning (Lambdin, 2003). There is, indeed, a need to equip teachers with a variety of open-ended problems for teaching through problem solving that can enhance their teaching options. Pupils must experience stimulating open-ended problems where they can reason and make their thinking visible. Pupils should also communicate their mathematical knowledge and find connections across mathematics as well as learn new mathematical concepts through an open-ended problem. Hence, this chapter reviews the notion of teaching through problem solving, discusses teaching through problem solving by making connections and illustrates lesson structure for teaching through problem solving.

2 Review of Literature

This section explains the notion of teaching through problem solving, and discusses making connections in relation to teaching through problem solving. In addition, lesson structure for teaching through problem solving are also discussed.

2.1 *Teaching Through Problem Solving*

Schroeder and Lester (1989) analysed three approaches to teaching problem solving: teaching *about* problem solving, teaching *for* problem solving, and teaching *through* problem solving. Teaching about problem solving refers to explicitly teaching strategies one can use to solve problems effectively. The emphasis is on using strategies to approach and solve problem that is normally not domain-specific to any topics in the curriculum. Teaching for problem solving focuses on learning mathematics for the purpose of applying it to solve problems after learning a particular topic. Teaching through problem solving is a comparatively new conception in the history of problem solving in the mathematics curriculum (Lester, 1994). Connections with inquiry-based learning (IBL) can easily be identified in teaching through problem solving (Artigue & Blomhøj, 2013). "Rather than rely on the teacher as an unquestioned authority, students in [inquiry-based] classrooms are expected to propose and defend mathematical ideas and conjectures and to respond thoughtfully to the mathematical arguments of their peers" (Goos, 2004, p. 259). Pupils undertake a very self-motivated role in their learning.

In teaching through problem solving, Cai (2010) articulated that the development of problem-solving competences is not disjoint from domain-specific objectives in the curriculum. According to Cai, teaching through problem solving starts with exploring a problem that is likely to offer students the opportunity to learn and understand important aspects of a mathematical concept or idea. Teaching through problem solving begins by giving pupils a problem, usually open-ended, to solve. Teaching through problem solving involves teaching new mathematical concepts through problem-solving contexts and enquiry-oriented environments that are planned by the mathematics teacher. This approach will create opportunities in "helping learners construct a deeper understanding of mathematical ideas and processes by engaging them in doing mathematics such as creating, conjecturing, exploring, testing, and verifying" (Lester et al., 1994, p. 154). Pupils would attempt to explore the problem situations and create their own solution representations with the guidance of the teachers through questioning, cues and hints. Pupils would also learn and understand critical aspects of the concept or idea by analysing the problem situation. Although less is known about the concrete plan pupils use to learn and make sense of mathematics through problem solving, there is consensus that teaching through problem solving holds the potential of promoting student learning (Schroeder & Lester, 1989). Many of the ideas typically associated with this approach (e.g., changing the teacher's roles, designing and choosing problems for teaching, collaborative learning) have been discussed widely.

In teaching through problem solving, two critical roles are expected from teachers to demonstrate in a classroom: (1) choosing suitable problems and (2) creating classroom discourse for fostering pupils' mathematical understanding (Cai, 2003). Teachers need to think about the types of open-ended tasks to pose; how to facilitate discourse in mathematics lesson, and how to support pupils' use of a variety of representations as tools for problem solving, reasoning, and communication.

2.2 *Pupils making connections in teaching through problem solving*

According to Coxford (1995), connection making is the ability to connect conceptual and procedural knowledge, using mathematics on other topics, using mathematics in real-world activities, using inter-topic connections in mathematics. In the context of the Singapore primary school mathematics curriculum, "connections refer to the ability to see and make linkages among mathematical ideas ..." (Ministry of Education, 2012, p. 17). "Making connections is at the heart of doing mathematics" (Hyde, 2007, p. 46). In teaching through problem solving, developing in students the ability to connect may be facilitated using open-ended problems. Non-routine mathematical problems that are cognitively demanding have the potential to provide intellectual contexts for pupils' mathematical development. Such problems can enhance pupils' conceptual understanding, foster their capability to reason and communicate mathematically, and capture their interests and curiosity (Cai, 2014; Hiebert, 2003; Marcus & Fey, 2003; Van de Walle, 2003). The type of problems that teachers choose to use in the mathematics classroom determines the content pupils learn and the opportunities to learn (Stein, Grover, & Henningsen, 1996). A number of research studies have provided a clear evidence to support the connection between the nature of problems and student learning (Cai, 2014; Hiebert & Wearne, 1993; Stein & Lane, 1996; Stein, Remillard, & Smith, 2007).

In teaching through problem solving, problems form the structural focus and motivation for pupils' learning, and serve as a vehicle for mathematical exploration. There is significant prominence placed on exploratory activities, observation and discovery, and trial and error. In teaching through problem solving, learning deepens during the process of trying to tackle problems in which relevant mathematics concepts and skills are embedded (Lester & Charles, 2003; Schoen & Charles, 2003). As pupils solve problems, they can use any method they can think of, activate their prior knowledge or they can construct knowledge on the spot, and justify their ideas in ways they feel are convincing. In other words, it requires pupils to analyse an open-ended problem and then make connections by developing from previous concepts. Powerfully, teaching mathematics through problem solving enables pupils go beyond acquiring isolated ideas toward developing increasingly connected and complex system of knowledge (Cai, 2003; Hiebert & Wearne, 1993; Lambdin, 2003).

In teaching through problem solving, pupils must be able to solve the problem using previously learned concepts and skills and see connections among concepts rather than treating them as isolated ideas. The pupil's thinking and work on the problem also assist him or her make connections across different modes of external representations. Lesh, Post and Behr (1987) stressed that a pupil who "understands" a mathematical concept "can (1) recognise the idea embedded in a variety of qualitatively different representational systems, (2) flexibly manipulate the idea within given representational systems, and (3) accurately translate the idea from one system to another" (p. 36). Moreover, in this genuine engagement in problems, pupils are also believed to gain a good understanding of "intra-mathematical connections which link new mathematical knowledge with old, shaping it into a part of the mathematical system" (Noss, Healy, & Hoyles, 1997, p. 203).

Hiebert and Wearne (1993) in their study found that classrooms that emphasize teaching through problem solving used fewer problems and spent more time on each of them. They also pointed out that in a classroom using a problem-solving approach, teachers ask more conceptually-oriented questions (example: describe a strategy or explain underlying reasoning for getting an answer) and fewer recall questions than teachers in classrooms without a primary focus on problem solving. The study by Hiebert and Wearne (1993) showed that teachers should ensure a thoughtful use of time for effective organisation of problem-solving activities. Pupils' real opportunities to learn depend not only on the type of mathematical problems that teachers pose, but also on the kinds of classroom discourse that takes place during problem solving, both between the teacher and pupils and among pupils; frequently, teachers do not allow pupils to have productive struggle with cognitively demanding problems (Cazden, 1986). Discourse refers to the means of representing, thinking, discussing as well as agreeing and disagreeing that teachers and pupils use to engage in instructional tasks. The problem is solved by the pupils (or by groups of pupils), and the solutions are shared with the class.

Teaching pupils the solution is easy but encouraging pupils to create and improve on their ideas, and then seeing how all these connect to each other is a different experience altogether. The teacher needs to help pupils identify these different kinds of connections and build bridges across contexts to help pupils generalise their understanding. This could be implemented after individual attempts under teacher's guidance where the emphasis is put on collective sharing and discussion of different attempts and solutions. It also allows pupils to discover and discuss alternative approaches and solutions, and to clarify their own ideas. Tripathi (2008) contends that the role of the teacher in this process is to act as a facilitator "by asking questions that help students to review their knowledge and construct new connections" (p. 168).

Moreover, the learning environment of teaching through problem solving provides a natural setting for pupils to present various solutions to their group or class and learn mathematical concept through social interactions, meaning negotiation, and reaching shared understanding. Through presentation of various solutions, teachers should advocate pupils to examine how new mathematics concepts are connected to prior knowledge and experience, how new mathematics concepts are connected to real-world situations, how new mathematics concepts are connected within and between branches of mathematics and how the new mathematics concepts are connected to procedures. Lastly, teachers make concise summaries and lead pupils to understand important parts of the concept based on the problem and its multiple solutions. Such an instructional approach is more likely to enhance pupils' conceptual understanding and connections (Schroeder & Lester, 1989).

3 Lesson Structure for Teaching Through Problem-Solving

Teaching through problem solving is a pedagogy to engage pupils in problem solving in order to facilitate pupils' learning of important mathematical concepts and practices. The enactment of teaching through problem solving in the classroom is different from the traditional mathematics lessons where the objective is acquisition of mathematical concepts or skills by whole-class expository style. Problem solving lessons give the impression of being unstructured and hence difficult to assess whether students have achieved beyond solving the problems. Studies by Sullivan, Walker, Borcek, & Rennie (2015) and Bailey (2018) showed that it was necessary to have four phases of systematic lesson structure that foster problem solving and reasoning.

This chapter aims to provide teachers with a clearer guideline to enact teaching through problem solving. The lesson structure from Charles and Lester (1982), Sullivan et al. (2015) and Van de Walle (2007) which comprise four phases (*problem presentation* phase, *exploration* phase, *discussion of solutions and summary* phase and *consolidation* phase) is adapted in the discussion in this chapter. Each phase plays a vital role in the process of solving problem. The four phases change the teacher's mathematical practices and the student's behaviour and thinking in some ways.

3.1 *Four phases in teaching through problem solving*

The *problem presentation* phase refers to the time where the teacher presents the open-ended problem to the whole class and answered questions to help pupils clarify and understand the problem. The teacher focuses on essential information in the problem and retain the cognitive demands of the problem so that the problem is interpreted correctly. In order for pupils to experience productive struggle in generating understanding, the problem should not be simplified.

The *exploration* phase refers to the time pupils are working on the solution individually or in small group by exploring the use of strategies and heuristics. The teacher must allow time for pupils to explore the openness of the problem, generate conjectures, and build understanding. To help pupils clear blockages in solving the problem, the teacher could pose questions to refocus pupils' attention and give hints and cues only as a last resort. The teacher should anticipate how different pupils might respond to the challenge by appropriate scaffold, e.g. hints and cues that differentiate the experience (Bailey, 2018). At this *exploration* phase, the teacher is facilitator of learning through inquiry.

The *discussion of solutions and summary* phase, refers to the time the whole class discuss the solution attempts. Different groups of pupils should present their solution representations on the whiteboard illustrating different strategies used to find a solution. The display of the solution representations discloses how pupils process a problem and reveals the ways pupils connect their mathematical ideas and thinking processes. Sullivan et al. (2015), citing the work of Smith and Stein (2011), described the important features of *discussion of solutions and summary* phase as:

- Selecting responses for presentation to the class and giving those students some notice that they will be asked to explain what they have done;
- Sequencing those responses so that the reporting is cumulative; and
- Connecting the various strategies together (p. 45).

The reason for carefully selecting and sequencing responses is to lay the foundation for the teacher to connect the different solution representations to main mathematical ideas. By guiding pupils to focus on the connections between strategies and by shifting their focus from solution representations to mathematical ideas, teachers can commence to support pupils' efforts in understanding the new mathematical concepts based on the lesson objectives (Smith & Stein, 2011; Stein, Engle, Smith, & Hughes, 2008). In teaching through problem solving, it is important for pupils to actively construct the new mathematical concepts together with the teacher at this *discussion of solutions and summary* phase. Furthermore, the teacher can also highlight the different operative procedures and properties appropriate for the new mathematical concepts.

Finally, the *consolidation* phase refers to the time where further learning experiences are posed to consolidate the learning activated by the initial open-ended problem. The consolidation phase allows the teacher to provide opportunities for pupils to practise posing problems similar in structure and difficulty to the original open-ended problem. Some features of the original problem remain the same while other modifications to assist the pupil avoid over generalisation from solutions to one example.

3.2 *An illustration: Teaching Through Problem Solving*

A sample lesson plan, involving four phases, for a primary 4 class on area and perimeter of a rectangle is presented. This sample lesson is provided to highlight the types of questions, comments and cues teacher might include in teaching through problem solving session. For this sample lesson plan, the pre-requisite knowledge, lesson objectives and an open-ended problem (Table 1) are as follows:

1. *Pre-requisite knowledge.*
   Pupils should be able to find the area and perimeter of a rectangle given the length and breadth.
2. *Lesson Objectives.*
   At the end of the lesson, pupils should be able to:
   - identify how perimeter and area are related
   - find one dimension of a rectangle given the other dimension and its area or perimeter and
   - apply multiplication and division concepts to find one dimension of a rectangle given its area or perimeter and the other dimension.

Table 1
An open-ended problem used for teaching through problem solving

An Open-Ended Problem	Instructions for Pupils
John has a piece of wire 20 cm long. He wants to bend into the shape of a rectangle. (a) What are some of the sizes of rectangles that he could make? (b) Are you able to find the largest possible area of the rectangle? Explain and show your working.	Think first for a few minutes before beginning to discuss with each other. Listen to each other and try to understand each other's way of reasoning. Solve the problems through discussing, calculating, and drawing. You may use the 1-cm square grid to help you. When you have reached one or more solutions, each member of your pair should be able to explain how you have reasoned.

Table 2
The problem presentation phase

Use a whole-class discussion about understanding the problem

1. What does John want to make?
2. What is the length of the wire?
3. What does 20 cm stand for?
4. (For pupils who cannot differentiate the difference between the area and the perimeter) What do you mean by the perimeter of a rectangle? What do you mean by area of a rectangle?
5. What are the properties of a rectangle?
6. How do you find the perimeter and area of a rectangle?
7. What are you asked to find?
8. (This will avoid questions later in the *exploration* phase.) Can you assume that the length and breadth are in whole numbers?

An outline of a lesson plan format incorporating these four phases is shown in Tables 2, 3, 4 and 5. A teacher, in writing the lesson plan, needs to include problem-specific comments, guiding questions, and hints to the open-ended problem. The eight questions in the *problem presentation* phase serve to assist pupils to clearly comprehend the problem; and to enable the teacher to identify the students' misconceptions and difficulties. As calculator is not allowed to use at the primary 4 level, question 8 is to help pupils to focus on finding the lengths and breadths using whole numbers less than 20 cm. Question 1, 2, and 3 focus the students on the information in the problems, while questions 4, 5, 6, 7 and 8 focus students on inferring from the information in the problem.

Table 3
The exploration phase

Provides guiding questions, hints and cues as needed

1. A table or organized list might help you keep track of what you are doing.
2. Drawing different rectangles that might help you to keep track using the 1-cm square grid.
3. (For pupil who have completed solving the problem)
   How do you know that you have found all the possible rectangles?
   Show me how you work out your answer. Explain to me what you are doing as you work out your answer.
4. Teacher give 20 cm wire to the pupil. How do you make a rectangle? Draw the rectangle. What can you say about the length and breadth of the rectangle? Which sides are equal to each other? If the length is 8 cm, what is the breadth?

(Note: Hint 4 is a "last resort" suggestion for pupils who do not appear to make any progress)

At the *exploration* phase, the teacher may encounter pupils who find areas by guess-and-check at first but then become more systematic in their approach. The hints and guiding questions in the *exploration* phase, given to pupils who need assistance, help pupils to stay focused on the information given in the open-ended problem. As the pupils organise their thinking and assume the lengths and breadths are in whole numbers, the teacher could encourage pupils to either set up a table as shown in Table 4, or to draw different rectangles using 1cm square grids.

In the *discussion of solutions and summary* phase, pupils will work as a whole class, discussing, justifying, and challenging various solution representations to the problem. Pupils present their solution representations on the whiteboard illustrating different strategies used to find a solution. The students' solutions, irrespective of the correctness, should be presented to all the learners for discussion. Table 4 shows the first two solution representations, of which the first is based on the idea of listing all possibilities systematically, beginning with the longest length possible, and the second is based on consideration of different rectangles with different dimensions using 1 cm square grid.

The teacher could encourage pupils to make statements about the relationship between perimeter and the rectangle with the largest area. By guiding the pupils to focus on the connections between strategies and by shifting their focus from solution representation to mathematical ideas, the teacher can lead pupils to discover that the length of one side of a rectangle can be found if the area or perimeter and the length of the other side are given.

Table 4

The discussion of solutions and summary phase

Connect different solution representations to new mathematical concepts.

First Solution Representation

Length (cm)	Breadth (cm)	Area (cm²)	Perimeter (cm)
9	1	9	20
8	2	16	20
7	3	21	20
6	4	24	20
5	5	25	20
4	6	24	20
3	7	21	20
2	8	16	20
1	9	9	20

Second Solution Representation

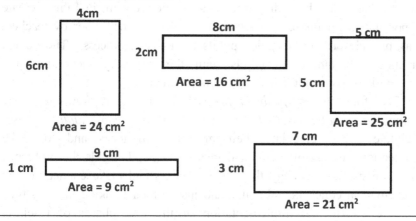

When the pupils are guided to discover that squares are special rectangles, they could easily connect the formulas for finding the area and perimeter of a square and a rectangle. Throughout the *discussion of solutions and summary* phase, the role of the teacher present the mathematical ideas in a connected and coherent manner rather isolated facts. In this case, pupils must able to explain how area of a rectangle, perimeter of a rectangle, length of a rectangle and breadth of a rectangle are all connected.

Table 5

The consolidation phase

Pose problems similar in structure and difficulty to the original open-ended problem
1.   Find the length and breadth of 2 different rectangles so that each has area of 14 cm^2.
2.   The perimeter of a rectangle is 18 cm. Find the possible area of the rectangle?

Classwork may be included in the *consolidation* phase. It is essential that pupils be given opportunity to internalise new concepts through applying it to the new problems. Table 5 shows two open-ended problems which the pupils are expected to apply non-algorithm thinking; to access relevant knowledge on area and perimeter of a rectangle; to apply area and perimeter concepts and to apply arithmetic manipulation skills to find the length and breadth of a rectangle. The problems in Table 5 create opportunities for pupils to have some guided practice whereby teachers may provide prompts to guide pupils through the process. The guided practice will familiarise pupils with the necessary techniques and expectations to accomplish independent practices.

The four phases (*problem presentation* phase, *exploration* phase, *discussion of solutions and summary* phase, and *consolidation* phase) facilitate pupils elucidate their mathematical ideas and to make connections among mathematical ideas. As pupils tackle the problems, they can use any strategy they know, access and make connections of mathematical concepts and skills, and justify their thinking in ways they feel are convincing. Implementation of the four phases of teaching through problem solving does not take much time from an already content-heavy curriculum, yet still allow teachers to emphasise the problem-solving process.

4 Concluding Remarks

In teaching through problem solving, teachers must move towards a more process-based approach in which getting a correct answer to a problem is not the only criterion. Teaching through problem solving can achieve both breadth and depth of learning, implying that the usual learning outcomes can still be achieved. It is imperative to note that teaching through problem solving means that pupils solve problem to learn new mathematical content, not just to apply mathematics after it has been learned. The lesson structure also offers a reflective framework against which the teachers could identify what needs to change and improve in their current problem-solving teaching practice. The emphasis is on leading pupils to discover the connections and guiding pupils to uncover these connections rather than teaching the content in isolation.

References

Artigue, M., & Blomhøj, M. (2013). Conceptualising inquiry-based education in mathematics. *ZDM-The International Journal on Mathematics Education, 45*(6), 797–810.

Bailey, J. (2018). Beginning teachers learning to teach mathematics through problem-solving. In Hunter, J., Perger, P., & Darragh, L. (Eds.). *Making waves, opening spaces (Proceedings of the 41st annual conference of the Mathematics Education Research Group of Australasia)* pp. 138-145. Auckland: MERGA.

Cai, J. (2003). What research tells us about teaching mathematics through problem solving. In F. Lester (Ed.), *Research and issues in teaching mathematics through problem solving* (pp. 241-254). Reston, VA: National Council of Teachers of Mathematics.

Cai, J. (2010). Commentary on problem solving heuristics, affect, and discrete mathematics: A representational discussion. In B. Sriraman & L. English (Eds.), *Theories of mathematics education* (pp. 251–258). New York: Springer.

Cai, J. (2014). Searching for evidence of curricular effect on the teaching and learning of mathematics: Some insights from the LieCal project. *Mathematics Education Research Journal, 26*, 811-831.

Cazden, C. B. (1986). Classroom discourse. In M. C. Wittrock (Ed.), *Handbook of research on teaching (3rd ed.)* (pp.432-463). New York: Macmillan.

Chang, S. C., Kaur, B., Koay, P. L., & Lee, N. H. (2001). An exploratory analysis of current pedagogical practices in primary mathematics classroom. *The NIE Researcher, 1*(2), 7-8.

Charles, R. I., & Lester, F. K. (1982). *Teaching problem solving: What, why and how.* Palo Alto, California: Dale Seymour.

Coxford, A. F. (1995). The case for connections. In P. A. House & A. F. Coxford (Eds), *Connecting mathematics across the curriculum* (pp 3–12). Reston: NCTM.

Goos, M. (2004). Learning mathematics in a classroom community of inquiry. *Journal for Research in Mathematics Education, 35*(4), 258-291.

Hiebert, J., Carpenter, T. P., Fennema, E., Fuson, K. C., Wearne, D., Murray, H., Olivier, A., & Human, P. (1997). *Making sense: Teaching and learning mathematics with understanding.* Portsmouth, NH: Heimann.

Hiebert, J. (2003). Signposts for teaching mathematics through problem solving. In F. K. Lester & R. I. Charles (Eds.), *Teaching mathematics through problem solving: Prekindergarten – grade 6* (pp. 53-61). Reston, VA: National Council of Teachers of Mathematics.

Hiebert, J., Gallimore, R., Gamier, H., Bogard G. K., Hollingsworth, H., Jacobs, J., Chui, A. M. Y., et al. (2003). *Teaching mathematics in seven countries: Results from the TIMSS 1999 video study.* Washington, DC: U.S. Department of Education, National Center for Educational Statistics.

Hiebert, J., & Wearne D. (1993). Instructional task, classroom discourse, and learning in second grade arithmetic. *American Educational Research Journal, 30*, 393-425.

Hiebert, J & Wearne, D. (2003). Developing Understanding through Problem Solving. . In L. S. Harold & R. I. Charles (Eds), *Teaching mathematics through problem solving: Grades 6–12* (pp. 3–12). Reston, VA: National Council of Teachers of Mathematics.

Hyde, A. (2007). Mathematics and Cognition. *Educational Leadership 65*(3), 43-47.

Lambdin, D. V. (2003). Benefits of teaching through problem solving. In F. K. Lester & R. I. Charles (Eds.), *Teaching mathematics through problem solving: Prekindergarten – grade 6* (pp. 3– 13). Reston, VA: National Council of Teachers of Mathematics.

Lesh, R., Post, T., & Behr, M. (1987). Representations and translations among representations in mathematics learning and problem solving. In C. Janvier (Ed.),

Problems of representation in the teaching and learning of mathematics (pp. 33–40). Hillsdale, NJ: Lawrence Erlbaum.

Lester, F. K. (1994). Musings about Mathematical Problem Solving Research: 1970-1994. *Journal for Research in Mathematics Education (25th anniversary special issue), 25* 660-675.

Lester, F. K.., Masingila, J. O., Mau, S. T., Lambdin, D. V., dos Santon, V. M. and Raymond, A. M. (1994). Learning how to teach via problem solving. In Aichele, D. and Coxford, A. (Eds.) *Professional development for teachers of mathematics* (pp. 152-166). Reston, Virginia: NCTM.

Lester, F. K. & Charles, R. I. (2003). *Teaching mathematics through problem solving: Pre-K–Grade 6*. Reston, VA: National Council of Teachers of Mathematics.

Mason, J., Burton, L., & Stacey, K. (1985). *Thinking Mathematically*. Wokingham, UK: Addison Wesley.

Marcus, R. & Fey, J. T. (2003). Selecting quality tasks for problem---based teaching. In H. L. Schoen & R. I. Charles (Eds.), *Teaching mathematics through problem solving: Grades 6-12* (pp. 55-67). Reston, VA: National Council of Teachers of Mathematics.

Ministry of Education (2012). *Primary mathematics: Teaching and learning syllabus.* Singapore: Curriculum Planning and Development Division.

Ministry of Education (2019). *Mathematics: Teaching and learning syllabus primary.* Singapore: Curriculum Planning and Development Division.

National Council of Teachers of Mathematics (NCTM). (2000). Principles and standards for school mathematics. Reston, VA: National Council of Teachers of Mathematics

Noss, R., Healy, L., & Hoyles, C. (1997). The construction of mathematical meanings: Connecting the visual with the symbolic. *Educational Studies in Mathematics, 33*(2), 203–233.

Schoenfeld, A. H. (1985). *Mathematical problem solving.* Orlando, FL: Academic Press.

Schoen, H. & Charles, R. (Ed.) (2003). Teaching mathematics through problem solving: Grades 6---12. Reston, VA: National Council of Teachers of Mathematics.

Schroeder, T. L., & Lester, F. K. (1989). Developing Understanding in Mathematics via Problem Solving. In P. R. Trafton & A P. Shulte (Eds.), *New directions for elementary school mathematics* (pp. 31–42). Reston, VA: National Council of Teachers of Mathematics, Inc.

Smith, M. S., & Stein, M. K. (2011). *5 practices for orchestrating productive mathematics discussions.* Reston, VA: National Council of Teachers of Mathematics.

Stein, M. K., Grover, B. W., & Henningsen, M. (1996). Building student capacity for mathematical thinking and reasoning: An analysis of mathematical tasks used in reform classrooms. *American Educational Research Journal, 33*(2), 455-488.

Stein, M. K., and Lane, S. (1996). Instructional tasks and the development of student capacity to think and reason: An analysis of the relationship between teaching and learning in a reform mathematics project. *Educational Research and Evaluation, 2*(1), 50-80.

Stein, M. K., Remillard, J., & Smith, M. S. (2007). How curriculum influences student learning. In F. K. Lester (Ed.), *Second handbook of research on mathematics teaching and learning* (pp. 319–369). Greenwich, CT: Information Age Publishing.

Stein, M. K., Engle, R. A., Smith, M. S., & Hughes, E. K. (2008) Orchestrating productive mathematical discussions: Five practices for helping teachers move beyond show and tell. *Mathematical Thinking and Learning, 10*(4), 313-340.

Sullivan, P., Walker, N., Borcek, C., & Rennie, M. (2015). Exploring a structure for mathematics lessons that foster problem solving and reasoning. In M. Marshman, V. Geiger & A. Bennison (Eds.). *Mathematics education in the margins* (Proceedings of the 38th annual conference of the Mathematics Education Research Group of Australasia), pp. 41-56. Sunshine Coast: MERGA.

Tripathi, P. (2008). Developing mathematical understanding through multiple representations. *Mathematics Teaching in the Middle School, 13*(8), 438-445.

Van de Walle, J. A. (2003). Designing and selecting problem---based tasks. In F. K. Lester & R. I. Charles (Eds.), *Teaching mathematics through problem solving: Prekindergarten – grade 6* (pp. 67-80). Reston, VA: National Council of Teachers of Mathematics.

Van de Walle, J. A. (2007). *Elementary and middle school mathematics: Teaching developmentally* (6th ed.) Boston, MA: Pearson/Allyn and Bacon.

Yeo, S. M., & Zhu, Y. (2005). Higher-order thinking in Singapore mathematics classrooms. *Proceedings of the international conference on education: Redesigning pedagogy: Research, policy, practice*. Singapore: Centre for Research in Pedagogy and Practice, National Institute of Education.

Chapter 8

Student-centred Learning Mathematics through the Lenses of the BSCS 5E Instructional Model

Diem H. Vuong

This chapter focuses on tailoring lesson plans toward student-centred instruction by implementing the Biological Sciences Curriculum Study (BSCS) 5E Instructional Model. Two exemplars of lessons for high school students using the BSCS 5E Instructional Model to allow students to make connections in solving real-world problems are presented.

1 What is BSCS 5E Instructional Model?

The Biological Sciences Curriculum Study (BSCS) 5E Instructional Model was originally designed for inquiry-based science lessons. The 5 "Es" in the BSCS Instructional Model are based on a recursive cycle framework of five cognitive stages in inquiry-based learning: *engagement, exploration, explanation, elaboration,* and *evaluation.* The objective of 5E Instructional Model in mathematics lessons is to provide students at any academic level to achieve the lesson's objectives and content mastery through making connections from prior knowledge and collaboration with peers in discovering and deriving new key concepts, theorems, and postulates.

The structure of 5E instructional model provides a lesson cycle in which direct teaching occurs concurrently with exploration activities, whereas the traditional inquiry-based learning project requires students to have more rigorous and comprehensive prior knowledge. The concept of the 5E Instructional Model is drawn from the learning philosophy of Johann Herbart (Bybee et al., 2006). The employment of the BSCS 5E Instructional Model promotes innovation in teaching and learning, where students have ownership of their learning by making connections to prior knowledge and discovering new learning objectives through problem-solving. The Nellie Mae Education Foundation demonstrates how a student-centred mathematics instruction can foster a mutual respect and trusting relationships between teachers and students (Walters et al., 2014). Within a safe learning environment, students engage in a more meaningful learning of mathematics and making connections to real-life application problems (see Figure 1).

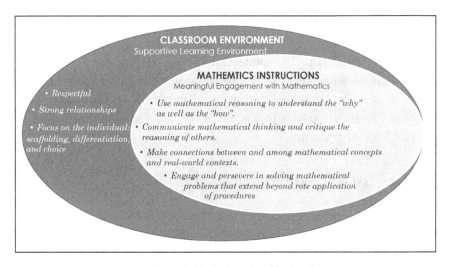

Figure 1. Student-centered instruction.
Resource: Nellie Mae Education Foundation, An Up-Close Look at Student-Centered Math Teaching.

2 Teaching Mathematics with BSCS 5E Instructional Model

Implementing the BSCS 5E Instructional Model in mathematics lessons can facilitate students in making connections between prior knowledge of previous lessons to real-world applications and new learning objectives. Within the BSCS 5E Instructional Model, teachers have three different roles: a *facilitator*, a *model*, and a *guide* to engage students in applying prerequisite concept skills to exploration activities and making connections to derive new concept skills.

As a facilitator, the teacher nurtures creative thinking, problem solving, interaction, communication, and discovery. As a model, the teacher initiates thinking processes, inspires positive attitudes toward learning, motivates, and demonstrates skill-building techniques. Finally, as a guide, the teacher helps to bridge language gaps and foster individuality, collaboration, and personal growth. The teacher flows in and out of these various roles within each lesson. (Instructional Strategies, 2015)

In the article "Mathematics through the 5E Instructional Model and Mathematical Modelling: The Geometrical Objects", Tezer & Cumhur (2017) defines Mathematics as follows:

Mathematics is a science, whose reflection we sometimes see directly in our lives and sometimes use to gain meaning in our lives. Therefore, Mathematics, which affects our lives, has great importance in our schools as a lesson. For this reason, it is necessary to carry out Mathematics courses in a way that will give students the ability to solve real-life problems. (Tezer & Cumhur, 2017, p. 4790)

Ebrahimi (2012) outlines the benefits of implementing the 5E Instructional Model in math lessons for the following reasons:
- Recalls students' prior knowledge.
- Encourages students to ask questions to promote content mastery.
- Engages students in mental and hands-on activities to make connections and explore to deepen understanding of the subject.

- Engages students in Writing-To-Learn (WTL)[1] activity to explain their conceptual knowledge on problem solving and mathematical procedures.
- Monitors and evaluates students' progress throughout the learning process.

The BCSC 5E Instructional Model method can be implemented along with the I Do, We Do, and You Do strategies. I Do is the initial phase of delivering a lesson, where the teacher performs direct teaching and models the lesson by connecting prior content knowledge skills to the new learning objectives. We Do is the second phase where students could collaborate with their classmates or to work together with the teacher in applying the newly learned skills as students are developing new mathematical content mastery. You Do is the final phase where students drill for skills independently as the teacher facilitates to check for students' understanding on the new lesson concepts (Lee, et al., 2020). In the remaining sections of this chapter, two exemplars of lessons in a typical high school that implement the BCSC 5E Instructional Model are presented.

3 Sample Lesson 1

In the first sample lesson for students in grade 9, the *learning objective* is that students will be able to solve simultaneous linear equations in two variables by substitution method. The *prior knowledge* required of the students is that they can solve linear equations with one variable.

---

[1] Writing-to-Learn activity promotes students critical thinking and recalling information that students learned. WTL maybe implemented in various ways, such as a daily 5-10 minutes journal at the end of each class by answering lesson's learning objectives. WTL activity also assists students learn new concepts better and retain information longer. *https://www.teachingchannel.org/video/writing-to-learn*

3.1 *Engagement*

The engagement phase begins with a 10-minute review of problems at the beginning of class to allow students to make connections between past and present learning experiences. The activity engages students in solving a real-life situation that can be represented by a simultaneous linear equation, which focus students' thinking on the learning outcomes of current activities.

Warm-Up Activity:

You are buying movie tickets online for your cousins. You must enter the number of tickets you wanted on the screen below. You have $68 on your PayPal account to spend. How many tickets can you buy?

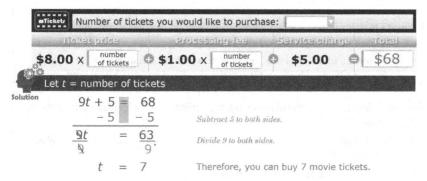

$$9t + 5 = 68$$
$$\underline{\quad -5 \quad \quad -5}$$ *Subtract 5 to both sides.*
$$\frac{9t}{9} = \frac{63}{9}.$$ *Divide 9 to both sides.*
$$t = 7$$ Therefore, you can buy 7 movie tickets.

Figure 2. The engagement phase of warm-up activity for Lesson 1

3.2 *Exploration*

The exploration phase provides students with a common base of experience in which they are familiar with or related to in their daily lives. They identify and develop concepts, processes, and skills. During this phase, they actively explore their environment with real-life connection problem or material manipulation (Figure 3). Teachers facilitate and guide students in problem solving through prompting questions. Potential responses from students are printed in *italic* fonts in Figure 3.

The flower shop allows you to trade flowers of equal value. The diagram shows two fair trades. The antique vase costs $169. What is the cost of the flower vase? Explain your reasoning.

Prompt Questions: Teacher facilitate students and guide students through prompt questions while students work in pairs or small groups.

- In the second trade, what can you substitute with each flower vase? [*The first trade shows that 1 flower vase is equal to 3 tulip plants and $40, so you may substitute this for each flower vase in trade 2.*]

- According to the second diagram, what combination of tulip plants and cash can you trade for an antique vase? [*7 tulip plants and $120*]

- Given that an antique vase is worth $169 cash, how much cash are 7 tulip plants cost? [*$49*]

- Hence, how much is the flower vase cost?
  [*$61, explanation may vary*]

Making Math Connection

Each trade can be modeled by an equation written in two variables. Simultaneously, the two equations can be solved to find a solution for each variable that satisfies both equations.

Figure 3. The exploration phase for Lesson 1

3.3 *Explanation*

In the explanation phase, the teacher guides while the students explain the concepts, they have been exploring. The students are given the opportunity to verbalize their conceptual understanding or to demonstrate new skills or behaviors. For example, the mathematical connections in the exploration activity of the flower vase activity in Figure 3 promotes minds-on activity, where students observe each piece of information can be modeled by equations written in two variables as follows:

Let t represents tulip plants and v represents flower vase.

Equation 1: $3f + 40 = v$

Equation 2: $2v + f + 40 = 169$

Thus, both equations have two unknown variables, this become simultaneous equations, which can be solved to find the solution for each unknown variable by substituting the expression, v, from equation 1 into equation 2 to solve for t.

Equation 2:	$2v + f + 40$	$=$	169	
	$2(3f + 40) + t + 40$	$=$	169	*(substitution)*
	$6f + 80 + t + 40$	$=$	169	*(distributive property)*
	$7t$	$=$	49	*(combined like-terms)*
	t	$=$	7	*(simplify)*
Equation 1:	$3t + 40$	$=$	v	
	$3(7) + 40$	$=$	v	*(substitution)*
	61	$=$	v	*(simplify)*

Hence, by substitution, the equations can be solved for each variable, i.e., the price of the tulip plants is $7 and the flower vase is $61.

The Explanation phase is comparable to the *I Do method*, where teacher models on making connections from prior content knowledge to the new learning objectives. This phase also provides the opportunity for the teacher to introduce formal terms, definitions, and explanations for concepts, processes, skills, or behaviors (Lee et al., 2020). Example 1 shows how the teacher works out the problem and explains the procedures of solving simultaneous equations as prescribed in the *I Do method*.

Example 1: Solve the simultaneous equations by the substitution method.

$$\begin{cases} 2x + 7 = -4 \\ x + y = -7 \end{cases}$$

Step 1: Solve for one variable in at least one equation, if necessary.		
$x + y = -7$	*Equation 2*	$x + y = -7$
$\underline{\quad -y \quad\quad -y}$		$\underline{\quad -x \quad\quad\quad -x}$
$x \quad\quad = -7 - y$	*Solve for one variable.*	$y = -7 - x$
Step 2: Substitute the resulting expression from step 1 into the other equation		
$2x + y = -4$	*Equation 1*	$2x + y = -4$
$2(-y - 7) + y = -4$		$2x + (-x - 7) = -4$
Step 3: Solve the equation from step 2 to get the value of the first variable.		
$2(-y - 7) + y = -4$	*Distributive Property*	$2x + (-x - 7) = -4$
$(-2y) - 14 \,(+\, y) = -4$		$(2x - x) - 7 = -4$
$-y - 14 = -4$	*Combine Like Terms*	$x - 7 = -4$
$-y = 10$	*Simplify*	$x = 3$
$y = -10$		
Step 4: Substitute the value in step 3 into one of the original equations and solve for the other variable.		
$x + y = -7$	*Equation 2*	$x + y = -7$
$x + (-10) = -7$	*Solve for the Variable*	$3 + y = -7$
$x - 10 = -7$		$y = -10$
$x = 3$		
Therefore, the solution is $(x, y) = (3, -10)$. Check the solution by substituting $(3, -10)$ into each of the original given equation.		

Figure 4. The explanation phase of Lesson 1.

3.4 *Elaboration*

The elaboration phase extends students' conceptual understanding and allows them to practice skills and behaviors. Through the new experience of expressing one variable in terms of the other as illustrated in Figure 1, the students develop deeper and broader understanding of major concepts, obtain more information about areas of interest, and refine their skills. Using the *We Do method,* the teacher guides students through prompt questions as in Example 2 (Figure 5) and allows students "thinking" time to work on the examples together with an elbow partner, classmates, or teacher. The *We Do method* provides opportunity for students to develop a deeper understanding of the learning objectives (Lee et al., 2020).

Example 2: Solve the system by substitution.

$$\begin{cases} y = 5x & (1) \\ x + y = -24 & (2) \end{cases}$$

Prompt Questions: Teacher model and guide students through problem solving.

- How can we get started? *If one equation is already given in one variable, use that equation for the substitution. If both given equations are already solved for a variable, we may use either equation.*

- What does a solution to both equations look like? *An ordered pair written as* (x, y).

- What does $x = -4$ represent? *The x-coordinate of the point where the two lines intersect.*

- What does $y = -20$ represent? *The y-coordinate of the point where the two lines intersect.*

- Suppose we solve the first equation for x. What expression would be substituted for x in the second equation? $\dfrac{y}{5}$

Example 3: Solve the system by substitution.

$$\begin{cases} 2y + 5x = -5 & (1) \\ -3x + y = -8 & (2) \end{cases}$$

Prompt Questions: Teacher guide students through problem solving.

- Think. Which variable should we solve first?
  The variable that has coefficient of either 1 or –1.

- Does it matter which variable we select to solve first?
  No. You may select/choose either variable to solve first.

- Which variable appears easiest to solve for? Explain. *Variable y in equation 2 appears to be easiest as its coefficient is 1.*

- If y in equation 2 was solved first, i.e. $y = 3x - 8$, then substitute $3x - 8$ into equation 1. What property do you use? Explain.
  Distributive Property to find the product of 2 and 3x − 8.

- Why can we substitute 1 for x in either equation?
  1 is the x-coordinate of a point that intersects on both lines.

Example 4: Real-World Connection

A juice bar sells two sizes of juice drinks. A medium is $4, and a large is $5. In one day, the juice bar sold 80 drinks for a total of $364. How many large size drink did the juice bar sell?

Write the simultaneous linear equations to represent/model the verbal descriptions in the word problem.

$$\begin{cases} x + y = 80 & (1) \\ 4x + 5y = 364 & (2) \end{cases}$$

Prompt Questions: Teacher facilitate and guide students through making connections to real-world applications and problem solving.

- Think. What does the solution represent in the real-world? *Depends on the assigned variables. For example, let **x** represents medium and **y** represents large. Thus, (36, 44) represents 36 medium size drinks and 44 large size drinks.*

- Why is it impossible for either coordinate to be a non-whole number? *It is not possible to sell part of a drink.*

Figure 5. The elaboration phase for Lesson 1

3.5 *Evaluation*

The evaluation phase encourages learners to assess their understanding and abilities and lets teachers evaluate students' understanding of key concepts and skill development. Evaluation phase of the 5E instructional model is like the *You Do* phase, where students may work independently, with a partner, or in small group of no more than 5 persons. Students at the evaluation phase have an active role in working out the problem or problems while the teacher circulates and checks for student understanding and mastery of the lesson objectives. The evaluation phase is especially encouraged for peer teaching or board work with only 5 to 10 minutes exit tickets (Lee, et al., 2020).

Closing: Teachers may close the lesson by engaging students with a journal writing or group discussion to describe a good situation for using the substitution method. (*Sample answer: A good situation occurs when one of the variables has a coefficient of 1.* Describe a difficult situation. *Sample: A difficult situation occurs when both equations have variables with coefficient greater than 1; i.e, neither equation can be easily solved for one of the variables.*)

4 Sample Lesson 2

In the first sample lesson for students in grade 10, the *learning objective* is that students will be able to find the exact value of the trigonometric ratios of the special right angles. The *prior knowledge* required of the students is that they can find an unknown measurement of one side of the right-angled triangle by applying Pythagoras' Theorem.

4.1 *Engagement*

The materials involved in the engagement activity of this lesson are square shaped and equilateral cut triangle origami papers with various measurements (i.e. squares with measurements on each side of 5 cm, 10 cm, and 15 cm and equilateral triangles with measurements on each side of 16 cm, 18 cm, and 20 cm), rulers, protractors, and pens or pencils. The purpose of this hands-on activity is to engage students in recalling key academic vocabulary, such as square, right angle, congruent sides, equilateral triangles, equiangular triangles, isosceles triangles, base angles, angle bisectors, diagonals, hypotenuse, altitude, perpendicular bisector, vertex of the triangle, and midpoint.

Each student is given three different measurements of the square shaped and an equilateral cut triangle origami paper. Teacher explains to students the following tasks:

- Measure the dimensions of one of the rectangular figures with a ruler and label the measurement of each side.
- Measure each angle with a protractor, label the measurement of each angle respectively, and identify the shape of the figure.

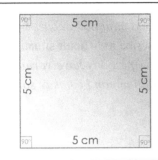

Recalling for Academic Vocabulary:

❑ Identify the given shape and explain your answer choice.

  ✓ Square

  ✓ Right Angle

  ✓ Congruent Sides

Figure 6a. The engagement phase for Lesson 2

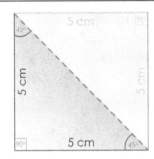

Recalling for Academic Vocabulary:

❑ Identify the given shape and explain your answer choice.

  ✓ Isosceles Triangle

  ✓ Base Angles

  ✓ Angle Bisectors

  ✓ Diagonal/Hypotenuse

Figure 6b. The engagement phase for Lesson 2

- Teacher asks students: "What shape is in Figure 6a? Explain your answer." [*The shape is a Square as the measures of each angle is 90 degrees, which is known as a right angle, and the measurement of each side are equals.*]

Fold along the diagonal of the square as illustrated in Figure 6b and identify the shape.

- Without using a ruler, calculate the exact length of the diagonal of the triangle, i.e. find the measure of c in figure 7.

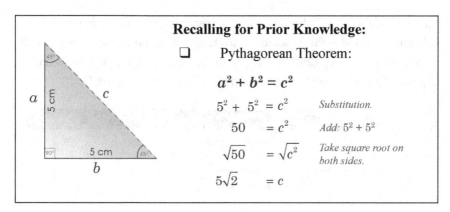

Figure 7. The engagement phase for Lesson 2

4.2 *Exploration*

During the exploration phase, students repeat the same protocols for square shapes with dimensions 10 cm x 10 cm and 15 cm x 15 cm to derive that the measurement of the hypotenuse of the $45° - 45° - 90°$ triangle is equal to the measurement of the side length times $\sqrt{2}$.

4.3 *Explanation*

Teachers explain to students by making math connection with another example as following: If the measures of the legs in a $45° - 45° - 90°$ triangle is *a*, what is the length of the hypotenuse (Figure 8)?

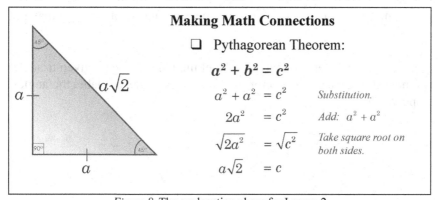

Figure 8. The explanation phase for Lesson 2

Students in this phase should be able to recognize the relationships of the side length of the right isosceles triangle and its hypotenuse. Teacher will then connect the exploration activity to the new learning objectives and explain the key concepts of trigonometric ratios as in Figure 9.

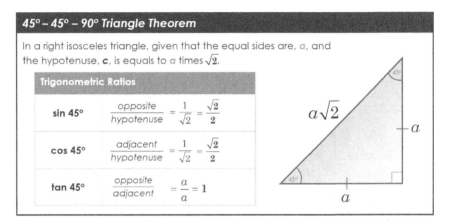

Figure 9. The explanation phase for Lesson 2

4.4 *Elaboration*

The lesson activity in the elaboration phase of this lesson is for students to discover and make mathematical connections in the $30° - 60° - 90°$ triangle. Teacher facilitates and guides students to derive the trigonometric ratios from various measurements of the equilateral triangles.

Step 1: Using a ruler, measure and label the length of each given triangle (Figure 10a). Using a protractor, measure each angle and label the angles respectively.

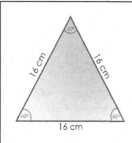

Recalling for Prior Knowledge:

❑ Identify the given shape and explain your answer choice.

✓ Equilateral Triangle

✓ Equiangular Triangle

Figure 10a. The elaboration phase for Lesson 2

<u>Step 2:</u> Fold along from one of the vertices to the midpoint of the opposite side of the triangle as illustrated in Figure 10b.

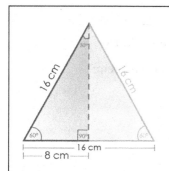

Recalling for Prior Knowledge:

❑ Identify the given shape and explain your answer choice.

✓ Right Triangle

✓ Altitude/Perpendicular Bisector

✓ Vertex/Vertices

✓ Midpoint

Figure 10b. The elaboration phase for Lesson 2

<u>Step 3:</u> Without using a ruler, calculate the exact length of the altitude or height of the triangle in Figure 11.

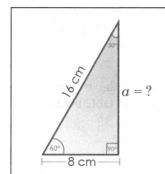

Recalling for Prior Knowledge:

❑ Pythagorean Theorem:

$$a^2 + b^2 = c^2$$

$$a^2 + 8^2 = 16^2 \quad \text{Substitution.}$$

$$a^2 = 192 \quad \text{Subtract : } 16^2 - 8^2$$

$$\sqrt{a^2} = \sqrt{192} \quad \text{Take square root on both sides.}$$

$$a = 8\sqrt{3}$$

Figure 11. The elaboration phase for Lesson 2

Step 4: If the hypotenuse of a $30^\circ - 60^\circ - 90^\circ$ triangle (Figure 12) is $2a$, what is the length of b (the shortest leg)? What is the length of h (the longest leg)?

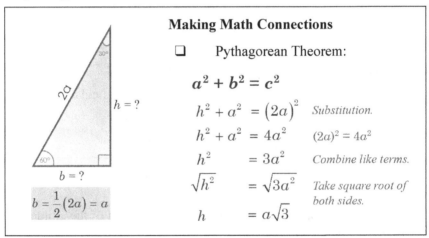

Making Math Connections

❑ Pythagorean Theorem:

$$a^2 + b^2 = c^2$$

$$h^2 + a^2 = (2a)^2 \quad \text{Substitution.}$$

$$h^2 + a^2 = 4a^2 \quad (2a)^2 = 4a^2$$

$$h^2 = 3a^2 \quad \text{Combine like terms.}$$

$$\sqrt{h^2} = \sqrt{3a^2} \quad \text{Take square root of both sides.}$$

$$h = a\sqrt{3}$$

In the triangle: $h = ?$, $b = ?$, $b = \frac{1}{2}(2a) = a$

Figure 12. The elaboration phase for Lesson 2

4.5 *Evaluation*

Students demonstrate lesson learning objectives of the trigonometric ratios of the $30^\circ - 60^\circ - 90^\circ$ triangle by completing the table below from the elaboration activity.

30° – 60° – 90° Triangle Theorem

In 30° – 60° – 90° triangle, the length of the hypotenuse, **c**, is $2a$. The length of the shortest leg is a, and the altitude measures $a\sqrt{3}$.

Trigonometric Ratios			
θ		30°	60°
sin θ	$\dfrac{opposite}{hypotenuse}$	$\dfrac{1}{2}$	$\dfrac{\sqrt{3}}{2}$
cos θ	$\dfrac{adjacent}{hypotenuse}$	$\dfrac{\sqrt{3}}{2}$	$\dfrac{1}{2}$
tan θ	$\dfrac{opposite}{adjacent}$	$\dfrac{1}{\sqrt{3}} = \dfrac{\sqrt{3}}{3}$	$\dfrac{\sqrt{3}}{1} = \sqrt{3}$

Figure 13. The evaluation phase for Lesson 2

Closing: In closing the Trigonometric Ratios lesson, students may recapture the key concepts by explaining how a right triangle is used to determine the exact value of sine, cosine, and tangent of an acute angle. Possible explanations on how a right triangle is applied in trigonometric ratio include *a right triangle is used to calculate sine by setting the ratio of the measurement of the opposite side to the measurement of the hypotenuse; cosine is the ratio of the measurement of the adjacent side to the measurement of the hypotenuse; and tangent is the ratio of the measurement of the opposite side to the measurement of the adjacent side.* Students would also be able to explain how to find the exact value of cos 65° without using a table or a calculator and the *sample answer is to sketch an acute right triangle with angle measurements of 25° – 65° – 90°. Measure and label each side length and the hypotenuse. Then, determine the trigonometric ratio of the measurement of the adjacent side length to the measurement of the hypotenuse.*

5 Summary of the BSCS 5E Instructional Model in Mathematics

The 5E Instructional Model consists of Engagement, Exploration, Explanation, Elaboration, and Evaluation. The first phase of the 5E is where the teachers engage students with *Do Now* problems that are related to the lesson's objectives or describe how the teacher will capture the student's interest. The second phase of the model is where the students explore the key concepts of the lesson through a *Discovery Activity* where teachers describe the hands-on or minds-on activities that the students will be doing and how students will be making connections to the new learning objectives. The third phase of the model is explanation through *direct instruction,* which covers key concepts, vocabulary, and problem solving. In the Explanation phase, teachers are given the opportunities to be creative to provide various instructional delivery techniques and innovation along with effective questioning strategies to help students connect their exploration to the concept under examination. Students are given time to drill for skills in the Elaboration phase of the 5E Instructional Model with *independent practice or small groups instructions* to gain knowledge on how to make real-world connections and apply in solving application problems. Evaluation is a closure in the 5E instructional model with the *exit ticket.* The exit ticket shows students' demonstration of understanding on the lesson that was taught.

6 Conclusion

The common core belief of mathematics educators is that mathematics instruction should focus on developing an understanding of concepts and procedures through problem-solving, reasoning, and discourse. In addition, students can learn mathematics through exploring and solving contextual mathematics problems. All students should have access to an array of strategies and approaches from which to choose, including general methods, standard algorithms, and procedures.

Although, the BSCS 5E Instructional Model is being implemented and promoted to be one of the instructional best practices in mathematics, its success depends on the teachers' belief and enactment of the lessons. Productive beliefs, such as seeing teaching mathematics as making connections between skills, properties, applications, and representations, are important in order to deliver effective instruction and achieve curriculum mastery. It is also crucial for teachers to consider how they can inject the joy of learning mathematics by engaging students in exploration activities as an alternative to simply teaching procedural and protocols for solving mathematical problems.

References

Bybee, R. W., et al. (2006). The BSCS 5E Instructional Model Origins, Effectiveness, and Applications. *http://www.bscs.org*. (Executive Summary)

Culver, Andrea (2019). Writing to Learn.
https://www.teachingchannel.org/video/writing-to-learn

Ebrahimi, S. (2012). Comparing the Effect of 5 E and Problem-Solving Teaching Methods on the Students' Educational Progress in the Experimental Sciences Course. *Journal of Basic and Applied Scientific Research.* 2(2)1091-1100.

Instructional Strategies Chapter of the Mathematics Framework (2015). The California Department of Education. *https://www.cde.ca.gov* (Curriculum Framework)

Lee, Maddie et al (2020). Strategies: I Do-We-Do-You-Do

https://strategiesforspecialinterventions.weebly.com/i-do-we-do-you-do.html

Walters, K. et al. (2014). An Up-Close Look at Student-Centered Math Teaching.

www.nmefoundation.org

Tezer, M., & Cumhur, M. (2017). Mathematics through the 5E Instructional Model and Mathematical Modelling: The Geometrical Objects. *EURASIA Journal of Mathematics Science and Technology Education.* 13(8):4789-4804. (Journal)

Chapter 9

Using Dialogue and Simulated Data in Teaching Probability

Von Bing YAP

A lesson plan for teaching expectation and standard deviation of the number of heads in several coin tosses is presented. It starts with a dialogue involving no technical terms, which motivates conjecture of their formulae and prepares for their interpretation to be revealed in an analysis of simulated coin tosses. The plan takes into account the fact that the expectation of a random variable is more intuitive that the variance or standard deviation.

1 Introduction

The axiomatisation of probability by Kolmogorov in 1933 is commonly viewed as the definitive step that firmly established probability as a branch of mathematics. However, the Kolmogorov axioms are indifferent to how probability is interpreted, hence are inadequate for guiding its application to data analysis problems, either to further scientific understanding or to make sound decisions. In my opinion, among the major interpretations, that based on frequency is the most important for applications. The frequency interpretation presupposes the existence of a random experiment which can be repeated indefinitely; this is very similar to the idea of a collective by Von Mises (1981), the first edition of which was in 1928. The probability of an event is then defined as the limit of the relative frequency of its occurrence as the number of repetitions goes to infinity. A finite number of repetitions affords an estimate of the probability, which becomes more precise as the number increases. The main strength of the frequency interpretation lies in its clarity and success in modelling a wide range of natural phenomena and human activities. For examples, see Freedman, Pisani and Purves (2007). A drawback is its silence on uncertain events which lie outside the notion of random experiments, including unique events. For instance, the probability that the average surface temperature of Earth will be 1 degree Celsius above the long-term average in year 2030 cannot be interpreted in terms of frequencies, unless one hypothesises a random mechanism that generates Earth-like planets indefinitely, which seems rather outlandish.

The Singapore O-level Mathematics introduces probability to 14-year-olds. The frequency interpretation is not explicitly stated, though reflected in some activities: "Compare and discuss the experimental and theoretical values of probability using computer simulations.". The aim of this chapter is to describe a lesson plan that features a dialogue, then an analysis of simulated coin tosses. Computer simulation has become indispensable in teaching probability and statistics in universities worldwide, due to the ease of obtaining large number of outcomes that look like they are generated from a specified random process. Indeed, simulation is set to significantly alter how statistical inference is taught, by enhancing or replacing some of the mathematical derivations that have been dominant in the textbooks. The dialogue is a novel feature, and it is different from the usual mathematical discourse, which helps students become familiar with technical terms and concepts after they have been taught. Rather, the proposed dialogue is designed to happen before explicit definitions are given. Indeed, a goal is to motivate the definitions through questioning and thinking about intuitive notions of chance which students already hold from life experience. Therefore there are some Socratic elements in the proposed dialogues. Elements of the lesson plan might already be well-known, given the rich literature on teaching probability. Interested readers may refer to books edited by Chernoff et al. (2016) and Kapadia & Borovcnik (1991).

2 The Dialogue

The main purpose of dialogues is to frame students' mind in the experimental context, so that they can reflect upon their intuitive notions. Prior opinions can be either affirmed as correct or modified. Technical terms should not be defined, in order for intuition to operate freely, and for the mind to have a more vivid impression when the definitions are presented.

2.1 *Expected Number*

The aim of this dialogue is to help students become familiar with the equivalent statements "The expected number of heads in n tosses of a coin with chance p of showing head is np." and "Let X be the number of heads in n tosses of a coin with chance p of showing head. The expectation of X is np." The priority is to make the statements intuitively convincing without going into their precise meaning, which will be dealt with in Section 3. Notice that the phrases "expected number" and "expectation" are introduced gradually towards the end of the dialogue.

Ask "Toss a fair coin 10 times. How many heads would you expect?". The most common answer is 5. This is not a precise question, since "expect" is undefined. Students may not be able to explain what 5 means, though they might agree that it does not mean 5 is always observed from the experiment. Next ask "Toss a fair coin 11 times. How many heads would you expect?". The form of the question tends to constrain the answer to be an integer. Most would offer 6, and some others would offer 5. To the student response "5 or 6", insist on a definite number. Then ask for a reason. One plausible reason for choosing 6 was that 11 was more than 10, which is not unreasonable. Students who offer 5.5 can be asked to think about what it means.

Now that the answer is either 5 or 6, there are two possible ways to continue. The first is to ask "Toss a fair coin 12 times. How many heads would you expect?". They should answer 6, and then will likely change the previous answer to 5.5. The second is to ask "Toss a fair coin 11 times. How many tails would you expect?". By symmetry, the answer should be the same as that for heads, but the sum of the two numbers will not be 11.

Even though students might agree that we should expect 5.5 heads in 11 tosses of a fair coin, they might find it a strange statement. After all, 5.5 is not a possible outcome of the experiment. This is a good opportunity to introduce the term "expected number":

The expected number of heads in 11 tosses is 5.5.

Similarly, in tossing 10 and 12 times, the expected numbers of heads are 5 and 6 respectively. Ask "What is the expected number of heads in n tosses of a fair coin?". We may get the correct answer $n/2$, if necessary by prompting students to refer to the three examples. Finally, get students to think about a p-coin, which has probability p of showing head. Intuitively, if $p < 0.5$, then we expect repeated tosses of a p-coin to yield less heads than a fair coin; the opposite holds for $p > 0.5$. The answer to "What is the expected number of heads in n tosses of a p-coin?" should be np. If it is not forthcoming, one might need to go through the intermediate case of a specific value of n.

Finally, two more new terms can be introduced. In the experiment of making n tosses of a p-coin, the number of heads is a *random variable*. We say that the *expectation* of the number of heads is np, which means exactly the same as saying the expected number of heads is np.

2.2 *Probability*

The gambler's fallacy refers to the belief that if something has happened more frequently than normal, it will happen less frequently in the future, or vice versa. It is false if the outcomes are generated from independent repetitions of a random experiment, such as tossing a coin or spinning a roulette. The aim of this dialogue is to discuss an extreme form of the gambler's fallacy associated with the classical method of assigning probabilities, i.e., equally likely outcomes.

Consider rolling a fair die, which has a 1/6 chance of getting six spots. Ask students if they agree that "In 6 rolls of the die, six spots will be observed exactly once." Ask those who disagree to change the mind of those students who agree, or hypothetical students who agree. Next, ask them to imagine the first roll yields six spots, and to predict what would happen in the next 5 rolls. If the statement is true, each of the 5 rolls must not yield six spots, which is inconsistent with experience. After all, there is no reason for the subsequent rolls to "know" what has happened before. How about "In 12 rolls, six spots will be observed exactly twice."? If the first 2 rolls yield six spots, the statement implies that the next 10 rolls must all stay away from six spots: another absurdity. Let k be a large integer. Check that students disagree with "In $6k$ rolls, six spots will be observed exactly k times.". It is difficult to understand the chance 1/6 with a fixed number of repeated experiments this way. Mysteriously, if we imagine infinitely many repetitions, then by the frequency interpretation, we may say that exactly 1/6 of them will produce six spots. I use "may say", because there is no clear way to explain what 1/6 of infinity means. In conclusion, the gambler's fallacy ceases to be so when we contemplate an infinite number of repeated experiments: the situation is qualitatively different from that of a finite number of experiments.

Might those students who initially agree with the question above also have agreed that in 10 tosses of a fair coin, exactly 5 heads will be observed? Though mathematically equivalent, life experience may lead them to respond differently.

3 Exploration of Simulated Coin Tosses

Many computer programmes can simulate various probabilistic phenomena. The resulting data can be very helpful for building valuable intuitions about probability. Here we focus on using simulated coin tosses to understand the expectation and standard deviation of the number of heads as a random variable, before delving into the basic notion of probability.

3.1 *Expected Number, or Expectation*

Simulate 12 tosses of a fair coin as a sequence of 0's (for tail) and 1's (for head). Repeat this process many times, say 1,000 times. Organise the data in a matrix of 1,000 rows and 12 columns. The above can be implemented in the programming language R using the following commands, which stores the matrix in a CSV file:

```
toss = matrix(rbinom(12000, 1, 0.5), 1000, 12)

write.table(data.frame(toss), file="toss.csv", sep=",", col.names=F, row.names=F)
```

Call this Dataset 1. The data file can be explored in EXCEL where the 12 tosses occupy the first 12 columns (A to L). Use 12 other columns (say O to Z) to record the cumulative sums of the first few columns: Column O is identical to column A. Column P is the sum of columns A and B, etc. Column Z is the sum of columns A to L. Thus, columns X, Y and Z display the observed number of heads from 10, 11 and 12 tosses, repeated 1,000 times. The fluctuation within each column illustrates randomness, i.e., number of heads is a random variable. In particular, students can see clearly that, for instance, in 10 tosses, one does not always observe exactly 5 heads. The histogram can be used to emphasise this point, and also to give an idea of the distribution of the number of heads. For example, in 10 tosses it is much more likely to get 5 heads than 1 head.

Ask students among columns X, Y and Z, which has the lowest mean, and which has the highest. Next, ask them to guess the value of the three means, before and after showing a histogram in Figure 1. The area under a rectangle is numerically its height, since its width is 1, and it represents the proportion of the corresponding number. For example, the rectangle above 5 has a height of 0.25, so around 25% of the 1,000 numbers are 5. The total area under all rectangles is 1, or 100%.

1000 repetitions

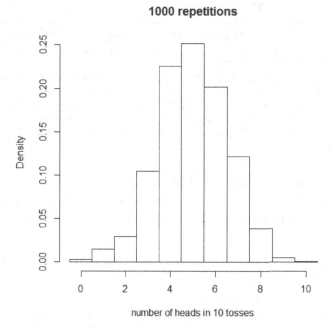

Figure 1. Histogram of column X from Dataset 1, consisting of 1,000 simulated number of heads in 10 tosses of a fair coin. Each rectangle is centred on an observed value, with width equal to 1, so its area is equal to the value of its height, or density, and this is also the proportion of that observation in the dataset.

The means of all 24 columns can be displayed in row 1001; of which the last 12 are collected in Figure 2 row "Mean". The means for X, Y and Z are close to 5, 5.5 and 6 respectively, which are in accord with estimates from histograms, and support the following interpretation of expected number. "In 11 tosses, the expected number of heads is 5.5" means: if the experiment is repeated many times to produce a long list of number of heads, then their average will be close to 5.5. The values in the row "Mean" increase by about 0.5 as we move one cell to the right, suggesting that the expectation of the number of heads in n tosses of a fair coin is $n/2$. Namely, if the experiment is repeated many times, the observed mean number of heads will be quite close to $n/2$. If Dataset 1 were regenerated, its content will be different, and hence the observed means will also be different, but the overall trend will still hold. This is confirmed by another replicate simulation summarised by row "Mean2".

Column	O	P	Q	R	S	T	U	V	W	X	Y	Z
Mean	0.50	1.00	1.49	1.99	2.49	3.01	3.51	4.02	4.49	4.99	5.48	5.97
Mean2	0.51	1.03	1.53	2.04	2.54	3.04	3.53	4.01	4.51	5.01	5.52	6.04

Figure 2. Dataset 1 records 12,000 simulated tosses of a fair coin in 1,000 rows and 12 columns. The row "Mean" shows the means of 1,000 observations from the number of heads in 1 toss (column O), 2 tosses (column P), etc. 12 tosses (column Z). The values are shown to two decimal places. The means are quite close to the expectation, which is number of tosses times half. A similar trend is shown in the row "Mean2" from a replicate simulation.

Figure 3 summarises Dataset 2, which is exactly like Dataset 1, except that the coin has a 0.2 chance of showing head. Here the means of the last 12 columns increase from 0.20 in steps of about 0.20.

Column	O	P	Q	R	S	T	U	V	W	X	Y	Z
Mean	0.20	0.41	0.61	0.79	1.00	1.22	1.41	1.63	1.84	2.04	2.26	2.46

Figure 3. Dataset 2 records 12,000 simulated tosses of a 0.2-coin in 1,000 rows and 12 columns. Each mean and variance are calculated from 500 simulated observations from the number of heads in 1 toss (column O), 2 tosses (column P), etc. 12 tosses (column Z). The values are shown to two decimal places. The means, reported to two decimal places, are quite close to the expectation, which is number of tosses times 0.2.

The datasets provide empirical support for the following interpretation of np as the expectation of the number of heads in n tosses of a p-coin. If the experiment is repeated many times, the observed mean number of heads will be quite close to np.

3.2 *An Interlude: The Standard Deviation*

This subsection presents a dialogue on the standard deviation (SD) based on Datasets 1 and 2, before calculating the SDs. It does not appear in the previous section because students lack intuition about the SD of a random variable, unlike its expectation. However, after the datasets are made, a meaningful dialogue can be built upon students' knowledge about the SD as a descriptive statistic.

Dialogue: In Dataset 1, columns X, Y and Z contain 1,000 observed number of heads from 10, 11 and 12 tosses of a fair coin. Which column has the largest SD? Which column has the smallest SD? Guessing from the relative ranges is a reasonable strategy. The students may then be asked to guess the values of the SDs, just to know that they do not have much experience to rely upon.

You are given a list of 0's and 1's, where exactly half are 1's. What do you think the mean and SD are? The SD is the square root of the variance, which is defined in general as follows. Obtain the deviations by subtracting the mean from every number of the list. Calculate the mean of the squares of the deviations. This is the variance of the list of numbers. The mean and SD are both 0.5.

You are given a list of n numbers consisting of k 1's and n-k 0's, where $k < n$. Denote the proportion of 1's, k/n, by p_1. Show that the mean of the list is p_1, and the variance is $p_1(1$-$p_1)$.

In Dataset 1, the mean of the first column is 0.498, so by the result in the previous paragraph, among the 1,000 observations of the first toss, the proportion of heads is 0.498. What is the SD of this column? Students' calculation can be confirmed in EXCEL using the command SD.P(A1:A1000).

Column X contains 1,000 observed number of heads in 10 tosses. What roughly is the variance, or the SD? Now we use the rule of thumb that in many histograms, around 68% of data are within 1 SD of the mean. Since the total area of the three central rectangles is around 67%, and their bases go from 3.5 to 6.5, we guess the SD is around 1.5. Then the variance is around the square of 1.5, or around 2.25. This ends the dialogue.

In Dataset 1, calculate the variance of all 24 columns in row 1002 using VAR.P(). VAR.S() computes the sample variance, which is slightly larger. Each of the first 12 cells in row 1001 holds the proportion of 1's in the column above. Call this number p_1, and it follows that the cell below should hold the value of $p_1(1$-$p_1)$. The first 12 values in row 1001 are all around 0.5, which is the expected number of heads in one toss, and those in row 1002 are all around 0.25=0.5(1-0.5). Now shift attention to the last 12 variances, which are collected in Figure 4. The variances increase from 0.25 by around 0.25 as we move from left to right, in the two replicates of Dataset 1. Moreover, the two variance for 10 tosses are quite close to our guess of 2.25 based on Figure 1. These

Mathematics - Connections and Beyond

observations support the conjecture that the variance of many observed number of heads in n tosses of a fair coin is around $n \times 0.5(1\text{-}0.5)$.

Column	O	P	Q	R	S	T	U	V	W	X	Y	Z
Var	0.25	0.48	0.70	0.97	1.16	1.40	1.65	1.92	2.16	2.38	2.62	2.97
Var2	0.25	0.48	0.76	1.00	1.23	1.50	1.71	1.91	2.20	2.36	2.62	2.81
Var0.2	0.16	0.32	0.48	0.64	0.80	0.99	1.14	1.28	1.44	1.60	1.77	1.94

Figure 4. Variances of 1,000 simulated observations from the number of heads in 1 toss (column O), 2 tosses (column P), etc. 12 tosses (column Z). The values are shown to two decimal places. The first two rows are from Dataset 1 (fair coin) and its replicate dataset,; the third row is for Dataset 2, using a 0.2-coin.

The last row of Figure 4 shows the variances for Dataset 2. The variances in the first 12 columns in row 1002 should all be around $0.16=0.2(1\text{-}0.2)$. The variances increase from 0.16 by around 0.16 as we move from left to right. These observations suggest that the variance of many observed number of heads in n tosses of a p-coin is around $np(1\text{-}p)$.

In summary, consider the experiment of tossing a p-coin n times. Based on the simulated data, we predict that if observations of number of heads are obtained from repeating the experiment many times, the data will have mean around np and variance around $np(1\text{-}p)$. Furthermore, the mean and variance will tend to be closer to the respective quantities with more observations. Just as we defined the expectation of *the number of heads* (a random variable) to be np, we now define its *variance* to be $np(1\text{-}p)$.

When students learn the general definitions of expectation and variance of a discrete random variable, and that the number of heads in n tosses of a p-coin has a binomial distribution, then the expectation np and variance $np(1\text{-}p)$ can be derived from the probabilities, rather than being directly defined as in the previous paragraph. Still, the interpretation is the same. For a random variable X, let its expectation and variance be denoted by $E(X)$ and $var(X)$ respectively. Obtain many observations from X. Their mean and variance will tend to be close to $E(X)$ and $var(X)$ respectively, and they tend to be closer with increasing number of observations. Let $SD(X)$ denote the standard deviation of X, i.e., the square root of $var(X)$. It is useful to describe a random variable X as follows: "*X* will be around E(*X*), give or take SD(*X*) or so." For example, in 100 tosses of a fair coin, the number of heads will be around 50, give or take 5 or so.

3.3 *Probability*

Finally, we use the simulated data to shed further light on the basic idea of probability. Imagine outcomes from tossing a fair coin arranged in a matrix with a number of rows and 10 columns. Let p_a be the mean of all the data, which is the proportion of 1's in all tosses, so is likely closer to 0.5 than any of the 10 column means, each being the proportion of 1's in only one-tenth tosses. But it can happen that one or more of the 10 proportions is closer to 0.5, highlighting the subtlety in the statement that a proportion from more observations tends to be closer to 0.5 than one from less observations. An example is shown in Figure 4, where the overall proportion of 0.5022 is beaten by the proportion 0.5020 in column D. However, we can argue that before we look at the 10 proportions, we have no idea which one will be closest to 0.5, so pitting p_a against the best of 10 proportions is unfair. A better way is to look at how far the 10 proportions are from 0.5. One can use the average of the absolute distances to 0.5, or the square root of the average of squared distances to 0.5. Both measures are likely larger than the distance between p_a and 0.5. An example is given in Figure 5.

Column	A	B	C	D	E	F	G	H	I	J
Proportion	0.498	0.502	0.490	0.498	0.502	0.516	0.504	0.507	0.476	0.494

Figure 5. Each proportion of 1's is calculated from 500 simulated tosses of a fair coin and is exact. The proportion in all 5,000 tosses is p_a=0.5022, which is also the average of the 10 proportions. The 4th proportion is closer to 0.5 than p_a. However, for the 10 proportions, their average absolute distance from 0.5 is 0.0258, and the square root of their average squared distance from 0.5 is 0.0313. By both measures, the 10 proportions are on the whole much further from 0.5 than p_a.

4 Conclusion

The dialogue on expected number of heads has been attempted on three occasions in 2019, to (i) about 30 science educators from ASEAN universities at a symposium in the National University of Singapore (NUS) in May, (ii) about 80 secondary school students at an outreach event by the NUS Department of Statistics and Applied Probability in June, and (iii) about 20 teachers at the Mathematics Teachers Conference. The last group also experienced the analysis of simulated data on expectation. Unfortunately, no feedback was solicited from any of the audience. Other parts of the lesson plan are yet to be tested in the field.

Once the random variable of interest X has been identified, the instructor has much control over the number of repetitions, i.e., the number of observed values to be generated from X. It is a good practice to use at least two different number of repetitions, in order to convince students that the larger the number of repetitions, the closer the mean and variance of the observed values tend to be to $E(X)$ and $var(X)$. This advice is not followed in Subsection 3.1, where the number of repetitions is kept at 1,000. However, in Subsection 3.3, comparisons are made between 500 and 5,000 repetitions to illustrate the point, with a caveat to look out for random surprise. We may prefer a larger number of repetitions, so that the observed means and variances are more stable. For instance, the trend in the variances in Figure 4 can become clearer with more repetitions. However, care should be taken so that students do not get the impression that with a "large enough" number of repetitions, the observed values will be equal to the target values. For example, with 20,000 tosses of a fair coin, the proportion of heads has over 0.9 chance of being equal to 0.50, if rounded to two decimal places. This means that if we simulate this proportion twice, we are quite likely to be struck that both values, to two decimal places, are exactly 0.50.

The dialogue may persuade many students that "How many heads would you expect in 11 tosses of a fair coin?" has a reasonable answer, namely 5.5. However, the instructor should refrain from insisting 5.5 as the correct answer. In my opinion, to hold the position that the question has a correct answer is overreaching, given that the word "expect" is not defined. In the dialogue, the strategy is to use a suitable moment to introduce the terms "expected number" and "expectation". It is wise to leave the dialogue with this question only slightly demystified, and largely retaining its vagueness from the beginning.

The frequency interpretation of probability is not intuitive, since it involves the imagination of potentially infinite repetitions of a random experiment. That most novices agree with or even make the statement "I expect 5 heads in 10 tosses of a fair coin.", or that perhaps some are even able to say "I expect 4 heads in 10 tosses of a 0.4-coin." might stem from a reasoning that seems to resemble the gambler's fallacy, that things even out. If the expectation of the number of heads is fairly intuitive, the same cannot be said of its variance. When told a fair coin is tossed 100 times, most people have no difficulty to expect 50 heads, but almost all will not be able to guess what the variance should be. The position of the dialogue on the SD in the lesson plan, after the analysis of means to cement the concept of expectation, reflects the higher challenge of grasping variance. The use of histograms supports the position that preparation in descriptive statistics helps students grasp the subtler concepts in probability.

References

Chernoff, E.J., Engel, J., Lee, H.S., & Sánchez, E. (Eds.) (2016). *Research on Teaching and Learning Probability*. Springer.

Freedman, D. A., Pisani, R. & Purves, R. (2007). *Statistics (4th ed)*. Norton.

Kapadia, R. & Borovcnik, M. (Eds.) (1991). *Chance Encounters: Probability in Education*. Springer.

Kolmogorov, A. N. (1933). *Grundbegriffe der Wahrscheinlichkeitsrechnung*. Berlin: Julius Springer. Translation: *Foundations of the Theory of Probability (2nd ed)* (1956). New York: Chelsea.

Von Mises, R. (1981) *Probability, Statistics, and Truth*. Dover.

Chapter 10

Unpacking the Big Idea of Equivalence

Joseph B. W. YEO

Equivalence is one of the eight clusters of big ideas proposed for the 2020 secondary school mathematics syllabus in Singapore. In this chapter, I will unpack the meaning of equivalence: the idea of an equivalence relation, the differences between equivalence and equality, and the usefulness of equivalent equations and equivalent statements in solving mathematical problems. In particular, I will examine the solutions of some types of equations which do not *seem* to produce equivalent equations in subsequent steps, e.g. the introduction of an extraneous solution when solving some equations involving surds or logarithms, and the elimination of a variable when solving a pair of simultaneous equations in two variables, and explain how the solutions of these equations can still produce equivalent equations in subsequent steps. In other words, the transformation or conversion from one equation to another equivalent equation is still the basis of the method for solving any kinds of equations. In addition, this chapter will discuss how to teach secondary school students the big idea of equivalence without teaching the abstract idea of an equivalence relation.

1 Introduction

Equivalence is a core idea in algebra (Knuth, Alibali, McNeil, Weinberg, & Madison, 2005). According to Rittle-Johnson, Matthews and Taylor (2011), equivalence in mathematics, "typically represented by the equal sign, is the principle that two sides of an equation represent the same value" (p. 85). However, how is equality the same or different from equivalence? For example, when solving an equation such as $3x + 5 = 2x + 12$, the expression $3x + 5$ is equal to the expression $2x + 12$ when $x = 7$, but is the expression $3x + 5$ equivalent to the expression $2x + 12$ as both expressions have the same value when $x = 7$? We also know that the relation 'is equal to' is just an example of an equivalence relation, which is taught only in university mathematics. Does it mean that we have to bring the idea of an equivalence relation down to the secondary school level?

Furthermore, when solving an equation such as $3x + 5 = 2x + 12$, we will manipulate the equation to form an equivalent equation, e.g. $x + 5 = 12$, whose solution set is the same as that of the original equation. However, during the solution of an equation such as $\sqrt{x^2 + 3} = 2x$, when we square both sides of the equation to obtain $x^2 + 3 = (2x)^2$, the resulting solutions $x = \pm 1$ contains an extraneous solution (i.e. an extra solution that is wrong, or not a solution of the original equation). The same applies to solving a pair of simultaneous equations in two variables. When we manipulate the two equations to form an equation in one variable by eliminating the other variable, the solution set of the equation in one variable is different from the solution set of the original pair of equations. Does that mean that it is not necessary to ensure that all the subsequent equations in the solution of an equation are equivalent? If the equations are not equivalent, i.e. they do not have the same solution set, then how can we be sure that the solution set of the last equation is, or contains, the solution set of the original equation?

Therefore, the purpose of this chapter is to unpack the big idea of equivalence so that teachers are clear about the meaning of equivalence and its relationship with the equal sign, and how to teach the concept of equivalence at the secondary school level without bringing in the more rigorous idea of equivalence as a relation. In particular, I will deal with the purpose of teaching equivalence in the first place, by looking at some important applications of equivalence in school mathematics. I will also deal with certain issues such as whether forming equivalent equations is the basis of solving *all* equations and why we end up introducing an extraneous solution in some cases.

2 Literature Review

Research (e.g. Behr, Erlwanger & Nichols, 1980; Kieran, 1981, 2006) has found that many students do not interpret the equal sign as what is on the right of the equal sign must have the same value as what is on the left of the equal sign. Instead, since primary school, many students have mistaken that the equal sign is a signal to 'do something in order to obtain an answer' because when they try to solve a question like $2 + 3 = ?$, their idea of the equal sign is that they have to do something (which is add 2 and 3 in this case) to obtain an answer and then write: $= 5$ (i.e. equal to the answer). Therefore, if they have to add 2 and 3 and then subtract 1, they will have no qualms writing it as: $2 + 3 = 5 - 1 = 4$, because they will interpret this as doing something to obtain 5, and then doing another thing to 5 in order to obtain 4.

To teach students the correct idea of equality, Kieran (1992) has suggested getting students to construct arithmetic equalities such as $2 \times 6 = 4 \times 3$, so that students will realise that what is on the right hand side (RHS) of the equal sign does not have to be a numerical value or answer, but it can also be an operation between two numbers, as long as it has the same value as the left hand side (LHS) of the equal sign.

Similarly, when it comes to algebra, "equations such as $3x + 5 = 26$ might fit the student's existing notions, but others such as $3x + 5 = 2x + 12$ might not" (Kieran, 1992, p. 399). This is because for the equation $3x + 5 = 26$, the right hand side of the equal sign is a numerical value, like an answer, but for the equation $3x + 5 = 2x + 12$, the right hand side of the equal sign is another algebraic expression.

Some researchers (e.g. Warren & Cooper, 2009) have suggested using a mathematical balance model to guide students to understand that the left hand side of the equal sign (or the balance) must be equal to the right hand side of the equal sign (or the balance). For example, as shown in Figure 1, the expression $3x + 5$ on the left hand side of the balance is equal to the expression $2x + 12$ on the right hand side of the balance. Therefore, if we subtract 5 from the left side of the balance so that the expression on the left is $3x$, we also have to subtract 5 from the right side of the balance so that the two expressions on the balance are still equal.

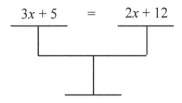

Figure 1. Balance model

However, some of these researchers *seem* to treat the equal sign in *any* equation as an equivalence. For example, Warren and Cooper (2009) started by saying that the equal sign in $3 + 4 = 7$ could be viewed as "equivalence or 'same value as'" (p. 76), which is similar to the definition of equivalence provided by Rittle-Johnson et al. (2011) as mentioned at the start of this chapter. Later in their paper, Warren and Cooper (2009) moved on to talk about using "balance representing equivalence" (p. 79) to teach equations, and that the balance model "considers both the right and left hand sides of equations" (p. 79). Finally, they mentioned that one key aspect of equivalence and equation is that the idea of equations as equivalence. Now, it *seems* that Warren and Cooper were saying that the left hand side of an equation is equivalent (rather than equal) to the right hand side of the equation, instead of using the idea of equivalent equations to solve an equation.

Similarly, Kieran (2006) *seems* to associate solving equations with producing equivalent expressions (instead of equivalent equations): "with respect to students' learning to solve equations and produce equivalent expressions" (p. 24). This *seems* to suggest the idea that the expressions on both sides must always be equivalent (rather than equal) when solving an equation.

For example, consider the equation $3x + 5 = 2x + 12$. The value of the expression on the left hand side when $x = 7$ (which is the solution) is equal to the value of the expression on the right hand side when $x = 7$. To solve the equation, we may first subtract 5 from both sides of the equation to obtain $3x = 2x + 7$. This second equation is equivalent to the original equation since both equations have the same solution set. In addition, for the second equation, the value of the expression on the left hand side when $x = 7$ (which is the solution) is still equal to the value of the expression on the right hand side when $x = 7$. But the two expressions are not equivalent because for two expressions in x to be equivalent, the two expressions must have the same value for *any* value of x. In other words, the basis of the procedure for solving equations is to convert the original equation to equivalent equations (and not equivalent expressions) because the two expressions on either side of the equation are not equivalent but they are just equal for that value of x which is the solution.

However, we also notice that the method for solving some equations does not *seem* to produce equivalent equations in subsequent steps of the solution. For instance, the equation in surds and the pair of simultaneous equations in two variables, as given in Section 1. This is a very important question because if subsequent equations are not equivalent to the original equation, then what is the basis of solving such equations? In this chapter, I will address some of these issues.

3 Equivalence as a Relation

For a more comprehensive understanding of equivalence, let us first look at the more rigorous idea of equivalence as a relation. A binary relation $\sim$ on a set X is said to be an *equivalence relation* if and only if it is reflexive, symmetric and transitive, i.e. for all a, b and c in X,

- $a \sim a$ (reflexive),
- $a \sim b$ if and only if $b \sim a$ (symmetric),
- if $a \sim b$ and $b \sim c$, then $a \sim c$ (transitive).

For example, the relation 'is equal to' (denoted by the symbol $=$) on a set of real numbers $\mathbb{R}$ is an equivalence relation because for all a, b and c in $\mathbb{R}$,

- $a = a$ (reflexive),
- $a = b$ if and only if $b = a$ (symmetric),
- if $a = b$ and $b = c$, then $a = c$ (transitive).

Another example of a equivalence relation is the relation 'is congruent to' (denoted by the symbol ≡) on the set T of all triangles because for all ΔA, ΔB and ΔC (where Δ is the symbol for triangle) in T,

- $\Delta A \equiv \Delta A$ (reflexive),
- $\Delta A \equiv \Delta B$ if and only if $\Delta B \equiv \Delta A$ (symmetric),
- if $\Delta A \equiv \Delta B$ and $\Delta B \equiv \Delta C$, then $\Delta A \equiv \Delta C$ (transitive).

Similarly, the relation 'is similar to' on the set T of all triangles is an equivalence relation because for all ΔA, ΔB and ΔC in T,

- ΔA is similar to ΔA (reflexive),
- ΔA is similar to ΔB if and only if ΔB is similar to ΔA (symmetric),
- if ΔA is similar to ΔB and ΔB is similar to ΔC, then ΔA is similar to ΔC (transitive).

To understand the concept of equivalence more deeply, we need to study some counter examples. A non-example of an equivalence relation is the relation 'is less than' (denoted by the symbol <) on the set of real numbers $\mathbb{R}$. Although the relation is transitive, it is not an equivalence relation because for all a, b and c in $\mathbb{R}$,

- $a \not< a$ (not reflexive),
- $a < b$ does not imply $b < a$ (not symmetric),
- if $a < b$ and $b < c$, then $a < c$ (transitive).

Although the relation 'is less than or equal to (denoted by the symbol $\leq$) on the set of real numbers $\mathbb{R}$ is reflexive ($a \leq a$ for all a in $\mathbb{R}$) and transitive (if $a \leq b$ and $b \leq c$, then $a \leq c$), it is still not symmetric ($a \leq b$ does not imply $b \leq a$).

Another counter example of an equivalence relation is the relation 'have a common factor greater than 1' between two positive integers that are greater than 1. Although the relation is reflexive and symmetric, it is not transitive, e.g. 3 and 12 have a common factor greater than 1, and 12 and 4 have a common factor greater than 1, but 3 and 4 do not have a common factor greater than 1.

From the above, we observe firstly that equivalence and equality are two different ideas (equality is just an example of an equivalence relation). Secondly, it is counter intuitive that 'is similar to' on the set of all triangles is an equivalence relation because the triangles are not congruent (or identical). In other words, equivalence is different from the idea of being exactly the same. Thirdly, teaching the definition of an equivalence relation to primary and secondary school students may not be age-appropriate. So the next question is how can we bring the idea of equivalence down to the level of these students?

Although the focus of this chapter is on developing the big idea of equivalence at the secondary school level, secondary school teachers need to know what Secondary 1 students have learnt about equivalence in primary school so as to build upon their prerequisite knowledge. Therefore, we will summarise what is taught at the primary level first.

4 Equivalence for Primary School Students

Although the implementation of big ideas in Singapore primary schools will only begin later with the 2021 Primary 1 students as the first cohort, the concepts of equivalent fractions and equivalent ratios are already in the current primary school mathematics syllabus. For example, in Primary 3, students learn about equivalent fractions, e.g. $\frac{1}{2}, \frac{2}{4}$ and $\frac{3}{6}$ are equivalent fractions; and in Primary 5, students learn about equivalent ratios, e.g. $1 : 2$, $2 : 4$ and $3 : 6$ are equivalent ratios. For a more detailed treatment of these concepts, the reader can refer to Yeo (2019).

However, one issue that primary school teachers and students face is when there is a remainder during division of whole numbers. For example, when 7 is divided by 3, the quotient is 2 and the remainder is 1. In all the three local primary school textbooks, this is written as $7 \div 3 = 2R1$. The question is whether $7 \div 3$ and $2R1$ are really equivalent since equality is an equivalence relation.

Now, $9 \div 4 = 2R1$ as well. Does that mean that $7 \div 3 = 2R1 = 9 \div 4$? But we know that $\frac{7}{3} \neq \frac{9}{4}$. On the other hand, $14 \div 6 = 2R2$ but $7 \div 3 = 2R1$. Yet $\frac{14}{6} = \frac{7}{3}$. The latter usually causes confusion for many students because they wrongly believe that $\frac{14}{6} \neq \frac{7}{3}$ because $2R2 \neq 2R1$.

Therefore, writing $7 \div 3 = 2R1$ is not correct because $7 \div 3$ and $2R1$ are not equivalent. Since students have also learnt in the textbooks that the quotient is 2 (other than the remainder is 1), they should just write, "When 7 is divided by 3, the quotient is 2 and the remainder is 1."

5 Equivalence for Secondary School Students

In secondary schools, teachers can build upon students' knowledge of equivalent fractions and equivalent ratios to expound on the big idea of equivalence, before proceeding to the equivalence of algebraic expressions, the equivalence of equations and the equivalence of mathematical statements.

5.1 *Equivalent fractions, decimals and ratios*

In Secondary 1, students will learn that a rational number is a number which can be expressed in the form of a fraction $\frac{a}{b}$, where a and b are integers such that $b \neq 0$. Therefore, teachers can take the opportunity to revisit the idea of equivalent fractions. In addition, students will also learn about real numbers which can be expressed as decimals, and in particular, rational numbers can be expressed as terminating or recurring decimals. In other words, a rational number such as $\frac{1}{2}$ is equal or equivalent to the decimal 0.5, and we write $\frac{1}{2} = 0.5$.

Later on, when students revisit the topic of ratios in Secondary 1, teachers can use the opportune moment to recap the idea of equivalent ratios. However, there lies a difficulty which students may encounter. For example, consider the following question:

If $x : 6 = 1 : 2$, find the value of x.

Since it is more cumbersome to manipulate two equivalent ratios, we can convert them to two equivalent fractions like this:

$$\frac{x}{6} = \frac{1}{2}.$$

Then we can easily solve for x. However, some students may develop the idea that $1 : 2 = \frac{1}{2}$, but this is not correct because $\frac{1}{2}$ is a number with a fixed value while a ratio is a comparison of two quantities that vary multiplicatively. Nevertheless, $x : 6 = 1 : 2$ is equivalent to $\frac{x}{6} = \frac{1}{2}$.

But so what if two fractions or two ratios are equivalent? What is the point of teaching students all these? The syllabus document stated, "The transformation or conversion from one form to another equivalent form is the basis of many manipulations for analysing and comparing them, as well as algorithms for finding solutions" (Ministry of Education of Singapore, 2018, p. S2-11). For example, is $\frac{2}{3}$ greater or smaller than $\frac{3}{4}$? These two fractions may not be easy for some students to compare. But if we convert $\frac{2}{3}$ and $\frac{3}{4}$ to their equivalent fractions of $\frac{8}{12}$ and $\frac{9}{12}$ respectively, we can easily observe that $\frac{2}{3}$ is smaller than $\frac{3}{4}$. Another example, as mentioned earlier, is the transformation of $x : 6 = 1 : 2$ into its equivalent equation $\frac{x}{6} = \frac{1}{2}$, which makes it easier for students to manipulate to solve for x.

Therefore, understanding the big idea of equivalence is not purely an academic pursuit but there are important implications in helping students perform various manipulations for analysis and comparison. In other words, teachers should make such applications explicit to students so that the latter could appreciate why they are learning the big idea of equivalence.

5.2 *Equivalent algebraic expressions*

An example that brings out the difference between equivalence and equality is algebraic expressions, which students will also learn in Secondary 1. For example, when expanding $2(x + 3)$ to become $2x + 6$, we will write $2(x + 3) = 2x + 6$. Now the two algebraic expressions $2(x + 3)$ and $2x + 6$ are not only equal but equivalent because the values of both expressions are equal for all values of x. In fact, $2(x + 3) = 2x + 6$ is an algebraic identity, albeit a trivial one, unlike the non-trivial identity $(a + b)^2 = a^2 + 2ab + b^2$. Although $2(x + 3) = 2x + 6$ may not be so helpful when viewed as an identity, the conversion from one form to another equivalent form is a useful manipulation when we want to, say, simplify an algebraic expression such as $2(x + 3) - x$:

$$2(x + 3) - x = 2x + 6 - x$$
$$= x + 6$$

But what about solving the equation $2(x + 3) = x + 7$? The two algebraic expressions $2(x + 3)$ and $x + 7$ are only equal for the solution $x = 1$, so they are not equivalent. Some students may compare $2(x + 3) = 2x + 6$ with $2(x + 3) = x + 7$ and observe that both are equalities, but only one of the two equalities is an equivalence relation, i.e. they may conclude that equivalence is a special case of equality when in fact, as discussed in Section 3, equality is an example of an equivalence relation. In other words, there is some kind of an overlap between equality and equivalence. What students have encountered in algebra so far are the equality of two expressions which may or may not be equivalent. So there is a need to let students experience for themselves examples of equivalence in algebra that are not equalities, e.g. equivalent equations, which students will encounter later when they solve linear equations in Secondary 1 (see Section 5.4).

Algebraic fraction vs. linear expression
Before we leave this section, let us consider another example when students have to simplify algebraic fractions in Secondary 2, such as the following:

$$\frac{x^2-1}{x-1} = \frac{(x+1)(x-1)}{x-1}$$
$$= x+1$$

The question is whether the original expression $\frac{x^2-1}{x-1}$ is equivalent to the last expression $x+1$. The issue here is that the original expression is not defined when $x=1$, but the last expression is defined when $x=1$. However, the purpose of simplifying an expression is to obtain a simpler equivalent expression. Therefore, if $\frac{x^2-1}{x-1}$ is not defined when $x=1$, then $x+1$ must also *not* be defined when $x=1$. In other words, the last expression $x+1$ has an unstated condition that $x \neq 1$, i.e. the last expression is actually $x+1$, where $x \neq 1$, which is not equivalent to the expression $x+1$, where x is any real number. Unfortunately, most school textbooks do not write the condition for an algebraic fraction to be defined as it is not in the syllabus.

To summarise, for this kind of manipulation where we divide a factor in the numerator of an algebraic fraction by the same factor in the denominator, it will still produce an equivalent expression if we take into account the domain for which the original expression is defined.

5.3 *Equivalent functions*

To view $\dfrac{x^2-1}{x-1}$ graphically, we can observe the graph of the function

defined by the equation $y=\dfrac{x^2-1}{x-1}$, where $x \neq 1$ (see Figure 2). Notice

that it is a straight line with the same equation as $y = x + 1$, except that there is a hole (denoted by an unshaded circle) when $x = 1$ because

$y=\dfrac{x^2-1}{x-1}$ is not defined at that value of x.

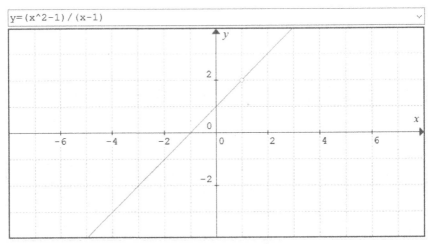

Figure 2. Graph of $y=\dfrac{x^2-1}{x-1}$

Therefore, the function $y=\dfrac{x^2-1}{x-1}$, where $x \neq 1$, is equivalent to the

function $y = x + 1$, where $x \neq 1$, because the two functions are identical or

exactly the same. But the function $y=\dfrac{x^2-1}{x-1}$, where $x \neq 1$, is *not*

equivalent to the function $y = x + 1$, where x is any real number, because their domains (i.e. values of x for which the functions are defined) are different. In other words, two functions are equivalent if they have the same equation, the same domain and the same co-domain (not just the same range).

Now, the manipulation that we have discussed so far is to divide the factor $x - 1$ in the numerator of the expression $\dfrac{x^2 - 1}{x - 1}$ by the same factor in the denominator. But what kind of manipulations can we perform on the equation of a function that will still produce an equivalent function and what kind of manipulations will not?

Linear law

Let us look at the Additional Mathematics topic of linear law, where students learn how to find the relationship between two variables when given some experimental data. For example, various experimental data are believed to obey the non-linear function $y = ax^2 + bx$. Since these are experimental data, they may not lie exactly on the curve $y = ax^2 + bx$, so we cannot just choose two data points and substitute into the equation of the curve to obtain two equations with the two unknowns a and b, and then solve for a and b simultaneously. Since it is not easy to draw a curve of best fit (unless one is using a graphing software to do this), one way to estimate the values of a and b is to first transform the equation of the non-linear function to the equation of a linear function:

$$y = ax^2 + bx$$

$$\frac{y}{x} = ax + b$$

$$Y = mX + c, \text{ where } Y = \frac{y}{x}, X = x, m = a \text{ and } c = b$$

The first question is whether the three functions above are equivalent. The issue here is that the first function $y = ax^2 + bx$ is defined for all real values of x, but the second function $\dfrac{y}{x} = ax + b$ is not defined when $x = 0$. Therefore, the first two functions are not equivalent because their domains are not the same, although their equations are the same.

However, if we use a graphing software to plot the graph of the second function $\frac{y}{x} = ax + b$ for some values of a and b, e.g. $\frac{y}{x} = 2x + 1$ as shown in Figure 3, the graph looks exactly like the graph of $y = 2x^2 + x$, which is defined when $x = 0$ because there is no hole at $x = 0$ (unlike the hole in the graph of $y = \frac{x^2 - 1}{x - 1}$ in Figure 2).

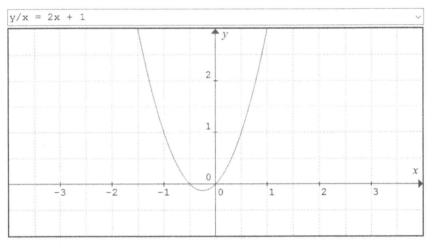

Figure 3. Graph of $\frac{y}{x} = 2x + 1$

This is because the software just interprets $\frac{y}{x} = ax + b$ as $y = ax^2 + bx$ in order to plot y against x, as indicated by the x- and y-axes. To verify that this is what the software does, let us express $y = \frac{x^2 - 1}{x - 1}$ as $y(x - 1) = x^2 - 1$ and plot the latter, which is shown in Figure 4. We observe that there is a hole at $x = 1$, but $y(x - 1) = x^2 - 1$ is defined when $x = 1$ because LHS = 0 = RHS when $x = 1$. Therefore, the software just interprets $y(x - 1) = x^2 - 1$ as $y = \frac{x^2 - 1}{x - 1}$, which is not defined when

$x = 1$. Similarly, the software just interprets $\dfrac{y}{x} = ax + b$, which is not defined when $x = 0$, as $y = ax^2 + bx$, which is defined when $x = 0$.

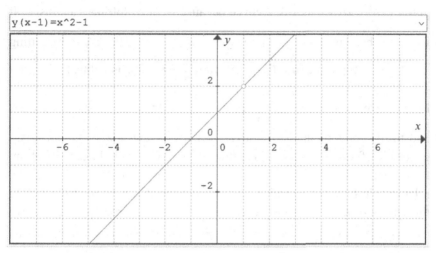

```
y(x-1)=x^2-1
```

Figure 4. Graph of $y(x - 1) = x^2 - 1$

Hence, it is crucial to know what a graphing software does in order to plot y against x so that we are aware that the software does not take into account the conditions for which the original function or equation is defined. For example, $\dfrac{y}{x} = ax + b$ is not defined when $x = 0$, but this particular software just plots $y = ax^2 + bx$, which is defined when $x = 0$.

Now, is the third function $Y = mX + c$, where $Y = \dfrac{y}{x}$, $X = x$, $m = a$ and $c = b$, equivalent to the second function $\dfrac{y}{x} = ax + b$? As we have observed earlier (see Figure 3), the graph of $\dfrac{y}{x} = ax + b$ is actually a curve and not a straight line. Because students are taught to express $y = ax^2 + bx$ as $\dfrac{y}{x} = ax + b$, in order to obtain a straight line graph, they may be confused that the graph of $\frac{y}{x} = ax + b$ is a straight line. Therefore, there is a need to guide students to see from Figure 3 that the graph of $\frac{y}{x} = ax + b$ is still a curve because we are plotting y against x. It will only become a straight line when we transform the curve by letting $Y = \frac{y}{x}$ and $X = x$, i.e. the third function $Y = mX + c$ is a linear function because we are plotting $Y = \frac{y}{x}$ against $X = x$. In other words, the second and third functions are not equivalent because the second function is a quadratic function while the third function is a linear function. Nevertheless, because of how we set up the transformation, there is still a relationship between m and c, and a and b, so that we can still estimate the values of a and b from the values of m and c. More specifically, we can say that the *non-linear form* $y = ax^2 + bx$ is transformed into its *equivalent linear form* $Y = mX + c$, where $Y = \frac{y}{x}$, $X = x$, $m = a$ and $c = b$. Therefore, converting a non-linear relationship between two variables into its equivalent linear form allows us to estimate the unknown parameters of the original function by determining the gradient and vertical intercept of the equivalent linear form of the function.

Before we leave this topic, there is another important issue to discuss about linear law. Students will be taught to use the experimental data to plot $Y = \frac{y}{x}$ against $X = x$, and then draw a line of best fit to find its gradient m and its Y-intercept c in order to obtain the values of a and b for $y = ax^2 + bx$ respectively. But why can we find the Y-intercept c since $Y = \frac{y}{x} = c$ is not defined at $x = 0$? This is because the limit of $\frac{y}{x}$ as x tends to 0 exists:

$$\lim_{x \to 0} \frac{y}{x} = \lim_{x \to 0} \frac{ax^2 + bx}{x}$$
$$= \lim_{x \to 0} (ax + b)$$
$$= b$$

Therefore, the Y-intercept c is equal to the limit b. This is the reason why we can still find the Y-intercept when it is not defined at $x = 0$. Since the idea of limits is not in the secondary school syllabus because it is beyond the cognitive level of most students, I suggest that we do not highlight this issue to the students, unless they ask. But for those who ask this kind of thought-provoking question, they are usually high progress learners, so the teacher can discuss this issue with them.

5.4 *Equivalent (conditional) equations*

Other than equations of functions discussed in Section 5.3, there are other types of equations:

- Conditional equations are equations that are true for some values of the variable(s), e.g. $2x + 1 = 7$;
- Contradiction equations are equations that are true for no values of the variable(s), e.g. $2x + 1 = 2x$;
- Identities are equations that are true for all values of the variable(s) for which they are defined, e.g. $2(x + 3) = 2x + 6$, $(a + b)^2 = a^2 + 2ab + b^2$.

In this section, we will focus on conditional equations (we will deal with identities in Section 5.5). Two conditional equations are equivalent if they have the same solution set, e.g. $2x + 1 = 7$ and $2x = 6$ are equivalent because they have the same solution set $x = 3$. We do *not* say that the two equations are equal and write $2x + 1 = 7 = 2x = 6$. So this is an example in algebra where we have an equivalence relation but not an equality. In this chapter, the phrase 'equivalent equations' will be used to mean 'equivalent conditional equations'.

An issue with teaching Secondary 1 students about the terminology 'the same solution set' is that they may not understand the word 'set': they may enquire why not just use the phrase 'the same solution'. The problem with the latter is that 'the same solution' can mean 'equal solutions', e.g. the quadratic equation $(x - 1)^2 = 0$ has two real and equal roots or solutions. This is different from the following two quadratic equations, $x^2 - 5x + 6 = 0$ and $(x - 2)(x - 3) = 0$, which are equivalent because they have the same solution set $\{2, 3\}$. But the only equations that students will encounter in Secondary 1 are linear equations, which only have one solution, so they may not appreciate what it means for an equation to have two solutions, not to mention two equal solutions. Therefore, for Secondary 1 students, we may have no choice but to say that two *linear* equations are equivalent if they have the same solution, and then extend the definition to 'the same solution set' only in Secondary 2 when they learn quadratic equations.

As mentioned earlier in Section 5.1, the syllabus document stated, "The transformation or conversion from one form to another equivalent form is the basis of many manipulations for analysing and comparing them, as well as algorithms for finding solutions" (Ministry of Education of Singapore, 2018, p. S2-11). For example, to solve $x^2 - 5x + 6 = 0$, we will transform the equation to the equivalent equation $(x - 2)(x - 3) = 0$, which is then equivalent to $x - 2 = 0$ or $x - 3 = 0$, and finally equivalent to $x = 2$ or $x = 3$. In other words, the basis of the procedure for solving an equation is convert the original equation into a series of equivalent equations. But as discussed in Section 1, we also notice that the method for solving some equations does not *seem* to produce equivalent equations in subsequent steps of the solution, which we will now examine in detail.

Solving equations involving surds

Let us consider the solution of the following equation in the Secondary 3 Additional Mathematics topic on surds:

$$\sqrt{x^2 + 3} = 2x$$
$$x^2 + 3 = (2x)^2$$
$$3x^2 = 3$$
$$x^2 = 1$$
$$x = \pm 1$$

Students are then taught that they have to reject the 'solution' $x = -1$ because LHS $= \sqrt{x^2 + 3} > 0$ but RHS $= 2x < 0$ if $x = -1$. However, if all the equations in the above working are equivalent, then why is there the extraneous solution $x = -1$ (an extraneous solution is an extra solution that is wrong)?

First, let us examine what produces the extraneous solution. In the original equation, we observe that $\sqrt{x^2 + 3} > 0$, so $2x > 0$, i.e. $x > 0$. But when we square $\sqrt{x^2 + 3}$ and $2x$ to become $\left(\sqrt{x^2 + 3}\right)^2$ and $(2x)^2$ respectively, we have added an extraneous solution in the form of $-\sqrt{x^2 + 3}$ (which is < 0) and $-2x$ (which is < 0) because $\left(-\sqrt{x^2 + 3}\right)^2 = \left(\sqrt{x^2 + 3}\right)^2$ and $(-2x)^2 = (2x)^2$. This is why $x = -1$ satisfies the second equation $x^2 + 3 = (2x)^2$ because $(-1)^2 + 3 = 4 = [2(-1)]^2$, but not the original equation $\sqrt{x^2 + 3} = 2x$. Hence, squaring an equation (or squaring the expression on either side of an equation) will introduce an extraneous solution, which is $x = -1$ in this example.

But does that mean the second equation $x^2 + 3 = (2x)^2$ is not equivalent to the original equation $\sqrt{x^2 + 3} = 2x$? Similar to what was discussed in Section 5.2 earlier that the algebraic fraction $\dfrac{x^2 - 1}{x - 1}$ has an unstated condition that $x \neq 1$, the original equation $\sqrt{x^2 + 3} = 2x$ also has an unstated condition: since $\sqrt{x^2 + 3} > 0$, then $2x > 0$ and so $x > 0$. Thus, if we present the working in the following manner, with the condition stated for each equation, then all the equations are equivalent:

$$\sqrt{x^2 + 3} = 2x, \qquad \text{where } x > 0$$
$$x^2 + 3 = (2x)^2, \quad \text{where } x > 0$$
$$3x^2 = 3, \qquad \text{where } x > 0$$
$$x^2 = 1, \qquad \text{where } x > 0$$
$$x = \pm 1, \qquad \text{where } x > 0$$
$$x = 1, \qquad \text{where } x > 0$$

Therefore, transforming the original equation into a series of equivalent equations is *still* the basis of solving the above equation involving surds. However, most of us would find it troublesome to write down the same condition for every equation. Instead, we would just check the 'solutions' $x = \pm 1$ at the last step to see which one satisfies the original equation and which one does not, in order to reject the extraneous one.

The next question is to find out if there are other manipulations that will result in an extraneous solution, and whether it is always easy to state the underlying condition of the original equation. To do these, we will look at the next three examples involving logarithms.

Solving logarithmic equations
Let us consider the solution of the following logarithmic equation in Secondary 3 Additional Mathematics:

$$2\lg x - \lg(x + 20) = 1$$
$$\lg x^2 - \lg(x + 20) = 1$$
$$\vdots$$
$$x = -10 \text{ or } 20$$

Students are then taught that they have to reject the 'solution' $x = -10$ because when $x = -10$, $2 \lg x = 2 \lg (-10)$ is not defined. First, let us examine what produces the extraneous solution. In the original equation, we observe that $x > 0$ for $2 \lg x$ to be defined. But when we convert $2 \lg x$ to $\lg x^2$, we have added an extraneous solution in the form of $-x$ (which is < 0) because $(-x)^2 = x^2$. This is why $x = -10$ satisfies the second equation $\lg x^2 - \lg(x + 20) = 1$ since $\lg(-10)^2 = \lg 100$ is defined, but not the original equation $2 \lg x - \lg(x + 20) = 1$ because $2 \lg(-10)$ is not defined. Hence, for this example, it is not the squaring of both sides of an equation but the squaring of the variable x when we convert $2 \lg x$ to $\lg x^2$ that introduces the extraneous solution $x = -10$.

But does that mean the second equation $\lg x^2 - \lg(x + 20) = 1$ is not equivalent to the original equation $2 \lg x - \lg(x + 20) = 1$? Similar to what was discussed earlier for the equation involving surds, the original equation $2 \lg x - \lg(x + 20) = 1$ also has an unstated condition: $x > 0$ for $2 \lg x$ to be defined, and $x + 20 > 0$ for $\lg(x + 20)$ to be defined. Solving both inequalities simultaneously, we obtain $x > 0$ and $x > -20$, which reduces to just $x > 0$, which is why we accept the solution $x = 20$ but not $x = -10$. Thus, if we present the working in the following manner, with the condition stated for each equation, then all the equations are equivalent:

$$2 \lg x - \lg(x + 20) = 1, \quad \text{where } x > 0$$
$$\lg x^2 - \lg(x + 20) = 1, \quad \text{where } x > 0$$
$$\vdots$$
$$x = -10 \text{ or } 20, \quad \text{where } x > 0$$
$$x = 20, \quad \text{where } x > 0$$

Therefore, transforming the original equation into a series of equivalent equations is *still* the basis of solving the above equation involving logarithms. But most of us would find it troublesome to write down the same condition for every equation. Instead, we would just check the 'solutions' $x = -10$ or 20 at the last step to see which one satisfies the original equation and which one does not, in order to reject the extraneous one.

Let us consider another example:

$$\log_3(x-5) + \log_3(x-4) = \log_3(x-1)$$
$$\log_3(x-5)(x-4) = \log_3(x-1)$$
$$\vdots$$
$$x = 3 \text{ or } 7$$

When $x = 3$, $\log_3(x-5) = \log_3(-2)$ and $\log_3(x-4) = \log_3(-1)$ are not defined. But when $x = 7$, $\log_3(x-5)$, $\log_3(x-4)$ and $\log_3(x-1)$ are all defined. However, we did not square the equation or any variable, so where does the extraneous solution come from?

In the original equation, we observe that $x - 5 > 0$ and $x - 4 > 0$ for $\log_3(x-5)$ and $\log_3(x-4)$ to be defined. But when we convert $\log_3(x-5) + \log_3(x-4)$ to $\log_3(x-5)(x-4)$, the conditions have changed to $(x-5)(x-4) > 0$ for $\log_3(x-5)(x-4)$ to be defined. But $(x-5)(x-4) > 0$ implies either (i) $x - 5 > 0$ and $x - 4 > 0$ (original conditions for x), or (ii) $x - 5 < 0$ and $x - 4 < 0$ (extraneous conditions for x). This is why $x = 3$ satisfies the extraneous conditions of $x - 5 < 0$ and $x - 4 < 0$, but not the original conditions of $x - 5 > 0$ and $x - 4 > 0$. Therefore, we observe that it is not just the squaring of an equation or a variable that can introduce an extraneous solution but taking the product of two positive algebraic expressions can also do so.

If we want to present the working so that all the equations are equivalent, we have to solve the three conditions $x - 5 > 0$, $x - 4 > 0$ and $x - 1 > 0$ simultaneously, which is not that difficult in this case. The end result is just $x > 5$, which explains why we accept $x = 7$ but not $x = 3$. But solving three inequalities simultaneously is not always so easy. For this, let us consider another example:

$$\log_3\left(x+4\right)+\log_3\left(2-x\right)=\log_3\left(2-3x\right)$$
$$\log_3\left(x+4\right)\left(2-x\right)=\log_3\left(2-3x\right)$$
$$\left(x+4\right)\left(2-x\right)=2-3x$$
$$\vdots$$
$$x=-2 \text{ or } 3$$

When $x = 3$, $\log_3\left(2-x\right)=\log_3\left(-1\right)$ and $\log_3\left(2-3x\right)=\log_3\left(-7\right)$ are undefined. But when $x = -2$, $\log_3\left(x+4\right)$, $\log_3\left(2-x\right)$ and $\log_3\left(2-3x\right)$ are all defined. However, if we want to present the working so that all the equations are equivalent, we have to solve the three conditions $x + 4 > 0$, $2 - x > 0$ and $2 - 3x > 0$ simultaneously, which is rather cumbersome in this case. This is another reason why we usually just solve the equation and then check the solution set of the subsequent equations to see which solution satisfies each of the three conditions of the original equation *separately* and which one does not.

Before we end this section, I would like to highlight that the reason why there is an extraneous solution for this last example is different from the previous example. For this last example, the first two equations are actually equivalent, unlike the previous example, because when $x = 3$, the second equation $\log_3\left(x+4\right)\left(2-x\right)=\log_3\left(2-3x\right)$ is still undefined, meaning that both the first two equations have the same solution $x = -2$ and so they are equivalent.

What is not equivalent is the third equation $\left(x+4\right)\left(2-x\right)=2-3x$ because its solution includes $x = 3$. So when we apply $\log_a x = \log_a y \Rightarrow x = y$, the underlying conditions that $x > 0$ and $y > 0$ (for $\log_a x = \log_a y$) have changed to include non-positive values of x and y (for $x = y$), thus resulting in the introduction of an extraneous solution.

To summarise, solving equations involving surds and logarithms *seems* to involve the conversion of the original equation into non-equivalent equations in some cases because of the introduction of an extraneous solution; and there are in fact quite a few types of manipulations (not just squaring) that can produce an extraneous solution. However, all the subsequent equations are actually equivalent to the original equation because the unstated underlying conditions for the original equation should also apply to all the subsequent equations. In other words, transforming the original equation into a series of equivalent equations is *still* the basis of solving the above equations involving surds or logarithms.

But what about solving simultaneous equations where it also seems to produce non-equivalent equations in subsequent steps?

Solving simultaneous equations
In Secondary 2, students learn how to solve a pair of linear equations simultaneously. For example, consider the following:

$$2x + y = 5 \text{ --- (1)}$$
$$x - y = 7 \text{ --- (2)}$$
$$(1) + (2): \quad 3x = 12$$
$$x = 4$$

Subst. $x = 4$ into (1): $\quad 2(4) + y = 5$
$$y = -3$$
$\therefore$ the solution is $x = 4$ and $y = 3$.

Now, the third equation $3x = 12$ has a different solution set from the original pair of equations. This is to be expected when we get rid of one unknown by adding the two original linear equations with two unknowns to obtain a linear equation with one unknown. However, the solution of the third equation, namely, $x = 4$, is still the solution of the unknown x in the original pair of linear equations. But why is this so if the equations are not equivalent? Moreover, why is the final solution $x = 4$ and $y = 3$ still the solution set of the original pair of equations if subsequent equations in the working are not equivalent?

The clue lies in substituting $x = 4$ into the first equation (or we can also substitute into the second equation) to obtain $y = 3$. What this means is that we should not view the third equation $3x = 12$ in isolation but together with the first (or the second) equation. In other words, when we add the first two equations (1) and (2), we actually convert the first pair of equations into another pair of equations (3) and (1) as shown below:

$$2x + y = 5 \text{ --- } (1)$$
$$x - y = 7 \text{ --- } (2)$$

$$(1) + (2): \quad 3x = 12 \text{ --- } (3)$$
$$2x + y = 5 \text{ --- } (1)$$

Now, the solution set of the second pair of equations (3) and (1) is the same as the solution set of the first pair of equations (1) and (2). This means that the second pair of equations is *still* equivalent to the first pair of equations. Simplifying equation (3), we will obtain the third equivalent pair of equations as shown below:

$$x = 4 \text{ --- } (4)$$
$$2x + y = 5 \text{ --- } (1)$$

Solving the third pair of equations by substituting (4) into (1), we will obtain the solution set:

$$x = 4 \text{ --- } (4)$$
$$y = -3 \text{ --- } (5)$$

Therefore, transforming the original pair of equations into a series of equivalent pairs of equations is *still* the basis of solving simultaneous equations, although in practice, we do not rewrite equation (1) so many times, but we only substitute (4) into (1) towards the end.

5.5 *Equivalent identities*

Let us now turn our attention to identities. Consider the following Pythagorean trigonometric identities:

$$\sin^2 \theta + \cos^2 \theta = 1$$
$$\tan^2 \theta + 1 = \sec^2 \theta$$
$$1 + \cot^2 \theta = \csc^2 \theta$$

If we divide the first identity $\sin^2\theta + \cos^2\theta = 1$ throughout by $\cos^2\theta$, we will obtain the second identity $\tan^2\theta + 1 = \sec^2\theta$; if we divide the first identity $\sin^2\theta + \cos^2\theta = 1$ throughout by $\sin^2\theta$, we will obtain the third identity $1 + \cot^2\theta = \text{cosec}^2\theta$. In other words, the three Pythagorean trigonometric identities are essentially the same.

But there is a difference: the conditions for the three identities to be defined are slightly different. Because we divide the first identity $\sin^2\theta + \cos^2\theta = 1$ throughout by $\cos^2\theta$ to obtain the second identity $\tan^2\theta + 1 = \sec^2\theta$, then $\cos^2\theta \neq 0$, i.e. $\theta \neq 90°, 270°$, etc. Therefore, the second identity $\tan^2\theta + 1 = \sec^2\theta$ is only defined when $\theta \neq 90°, 270°$, etc., which is true because $\tan\theta$ and $\sec\theta$ are not defined when $\theta = 90°$, $270°$, etc. (recall that an identity is an equation that is true for all values of the variable *for which it is defined*).

But what if we multiply the second identity $\tan^2\theta + 1 = \sec^2\theta$ throughout by $\cos^2\theta$ to obtain the first identity $\sin^2\theta + \cos^2\theta = 1$? Because $\sec^2\theta$ in the second identity is equal to $\dfrac{1}{\cos^2\theta}$, it is not defined when $\cos^2\theta = 0$. But when we multiply the second identity throughout by $\cos^2\theta$ to obtain the first identity, $\cos^2\theta$ can now be equal to 0, and so the first identity is defined even when $\cos^2\theta = 0$. The same can be said of $\tan^2\theta$ in the second identity which is not defined when $\theta = 90°$, $270°$, etc., but when $\tan^2\theta$ is multiplied by $\cos^2\theta$ to become $\sin^2\theta$ in the first identity, the latter is defined when $\theta = 90°, 270°$, etc.

Therefore, the three Pythagorean trigonometric identities are essentially the same except for the conditions for which they are defined. We can say that the three identities are equivalent.

There are various applications of equivalent identities. For example, depending on what is given, we can choose an appropriate Pythagorean trigonometric identity to apply. For instance, to solve the equation $2\cos^2 x + 3\sin x - 3 = 0$, we can use the first identity $\sin^2\theta + \cos^2\theta = 1$ (and not the other two identities) to convert $\cos^2 x$ to $\sin^2 x$ so as to form a quadratic equation in $\sin x$ to solve.

Another application of equivalent identities is in the proving of another identity. For example, to prove the following identity, one way is to start from the expression on the left hand side of the given identity and simplify it to become the expression on the right hand side.

$$\frac{\sin^2\theta + \cos^2\theta\sin^2\theta + \cos^4\theta}{\cos^2\theta} = \sec^2\theta$$

Another way (which is not the usual method taught in Singapore secondary school textbooks but is not wrong) is to convert the given equation (which is not an identity till proven) to a series of equivalent equations until we arrive at a known identity such as a Pythagorean trigonometric identity:

$$\frac{\sin^2\theta + \cos^2\theta\sin^2\theta + \cos^4\theta}{\cos^2\theta} = \sec^2\theta$$

$$\frac{\sin^2\theta + \cos^2\theta\left(\sin^2\theta + \cos^2\theta\right)}{\cos^2\theta} = \sec^2\theta$$

$$\frac{\sin^2\theta + \cos^2\theta}{\cos^2\theta} = \sec^2\theta$$

$$\tan^2\theta + 1 = \sec^2\theta$$

Since the last equation is a known identity, and all the equations are equivalent, then we have proven that the given equation is an identity. From another perspective, since all these equations are already proven to be identities, what we have is also a series of equivalent identities.

Partial fractions
The equivalence of identities seems to pose a thorny issue in Additional Mathematics when students learn how to decompose a proper algebraic fraction into partial fractions. Let us consider the following example:

$$\text{Let } \frac{5x+1}{(x-1)(x+2)} = \frac{A}{x-1} + \frac{B}{x+2}.$$

$$\text{Then} \qquad 5x+1 = A(x+2) + B(x-1).$$

$$\vdots$$

To find the value of A, we can substitute $x = 1$ into the second equation because the equation is an *identity*, i.e. it is true for any value of x. Similarly, to find the value of B, we can substitute $x = -2$ into the second identity. But the underlying conditions for the first identity to be defined are that $x \neq 1$ and $x \neq -2$. So, do these conditions apply to the second identity? If they do, how can we substitute $x = 1$ and $x = -2$ into the second equation to find the value of A and of B respectively?

Just like in Section 5.3 above when we were discussing about linear law, the key lies in the existence of the limits of $y = 5x + 1$ when $x = 1$ and $x = 2$. Graphically, the graph of $y = 5x + 1$, where $x \neq 1$ and $x \neq -2$, is a straight line with two holes at $x = 1$ and $x = -2$, as shown in Figure 5. Since the function $y = 5x + 1$ is continuous except for the two holes, the limit of $5x + 1$, as x tends to 1, exists, and is equal to 6, i.e. $\lim_{x \to 1}(5x+1) = 6$. Similarly, $\lim_{x \to -2}(5x+1) = -9$. Therefore, we are using these two limits to help us find the values of A and B in the second identity $5x + 1 = A(x + 2) + B(x - 1)$, even though the identity is not defined when $x = 1$ and $x = -2$, because the limits at these values of x exist.

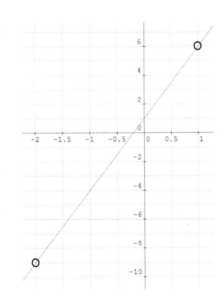

Figure 5. Graph of $y = 5x + 1$, where $x \neq 1$ and $x \neq -2$

As mentioned in Section 5.3, the idea of limits is not in the secondary school syllabus because it is beyond the cognitive level of most students. Therefore, I suggest that we also do not highlight this issue to the students, unless they ask. But, for those who ask, there is actually no need to bring in the idea of limit because the teacher can guide the students to observe from Figure 5 that when $x = 1$, $y = 6$, and when $x = -2$, $y = -9$, so we just use these observations to help us find the values of A and B in the second identity $5x + 1 = A(x + 2) + B(x - 1)$.

To summarise, there is a difference between equivalent (equations of) functions, equivalent (conditional) equations and equivalent identities. Nevertheless, the equivalence of all these equations is still the basis of algebraic manipulations for solving certain kinds of mathematical problems, such as solving (conditional) equations and proving identities.

5.6 *Equivalent mathematical statements*

The syllabus document (Ministry of Education of Singapore, 2018) also specifies the equivalence of mathematical statements: "When two mathematical statements are equivalent, it means that they imply each other." (p. S2-11). For example, two triangles are congruent if and only if all the corresponding sides of the two triangles are equal in length. Therefore, the statement 'two triangles are congruent' is equivalent to 'all the corresponding sides of the two triangles are equal in length' because they imply each other. An implication of the equivalence of these two statements is that we can prove that two triangles are congruent by proving that all the corresponding sides of the two triangles are equal in length.

Another example is the definition of a parallelogram: "A parallelogram is a quadrilateral in which opposite sides are parallel to each other". An equivalent statement is: "A parallelogram is a quadrilateral in which opposite sides are equal to each other." It can be proven that both statements imply each other and so are equivalent. An application of the equivalence of these two statements is that sometimes, based on what are given, it may be easier to prove that the opposite sides of a quadrilateral are equal (rather than parallel) to each other and so the quadrilateral is a parallelogram.

Therefore, equivalent mathematical statements are useful in solving certain kinds of mathematical problems, such as proving geometrical properties.

6 Implications for Teaching and Learning

To summarise, as the idea of an equivalence relation is beyond the cognitive level of most secondary school students, teachers can begin with the idea of equality and equivalence in arithmetic first, such as equivalent fractions and equivalent ratios, which students have learnt in primary school (see Sections 4 and 5.1).

Afterwards, when students encounter the equality of two algebraic expressions, we need to highlight that equality is different from equivalence: two algebraic expressions in x are equivalent if the values of both expressions are equal for any value of x, i.e. if the corresponding equation is an identity; but two algebraic expressions in x can be equal only for some values of x, in which case the equation is a conditional equation to be solved (see Section 5.2). At this stage, some students may think that equivalence is a special case of equality, so we can teach them that two linear equations are equivalent if they have the same solution (set) but we do not say that the two equations are equal. We have already discussed in Section 5.4 why Secondary 1 students may not understand the idea of 'the same solution set', so teachers can extend this idea in Secondary 2 when students encounter quadratic equations.

Also, in Secondary 1, teachers can impress upon the students that the equivalence of equations forms the basis of solving linear equations. But when they encounter solving a pair of linear equations simultaneously in Secondary 2, some students may wonder whether the equivalence of equations still forms the basis of solving such equations because after getting rid of one variable from a pair of linear equations in two variables, the resulting equation in one variable is not equivalent to the original pair of equations. The teacher can then use the opportune moment to guide students to realise that the resulting equation in one variable should not be viewed in isolation as explained in Section 5.4, but in fact the equivalence of equations still forms the basis of solving simultaneous equations.

In Secondary 3 Additional Mathematics, when students learn how to solve equations involving surds or logarithms, they will encounter algebraic manipulations that introduce an extraneous solution and some students may then wonder whether the equivalence of equations still forms the basis of solving such equations. The teacher can then use the opportune moment to guide students to realise that they have to take into account the unspoken conditions of the original equation when solving such equation, which will be used to get rid of the extraneous solution as explained in Section 5.4. Therefore, the equivalence of equations still forms the basis of solving such equations involving surds or logarithms.

As for the issue in Linear Law and Partial Fractions discussed in Sections 5.3 and 5.5 respectively, most students will not spot the problem. But those who do perceive the issue are usually high progress learners and the teacher can choose to explain to these students using the idea of limit, or just by looking at the graph without using the concept of limit, as explained in Sections 5.3 and 5.5.

Students also need to know that equivalence is not confined to mathematical objects but it can also apply to mathematical statements, which are useful to prove certain geometrical properties (see Section 5.6).

7 Conclusion

To conclude, the formal idea of equivalence as a relation is not an easy concept for secondary school students. But teaching students that equivalence is like 'equality' is a challenge because equivalence is different from equality although they are related concepts. Therefore, there is a need to highlight to students, at opportune moments, examples of equivalence that are not equality, and examples of equality that are not equivalence. More importantly, teachers should impress upon their students the usefulness of equivalence in solving certain mathematical problems at appropriate junctures so that their students will appreciate why equivalence is a big idea in mathematics.

References

Behr, M., Erlwanger, S., & Nichols, E. (1980). How children view the equals sign. *Mathematics Teaching, 92*, 13-18.

Kieran, C. (1981). Concepts associated with the equality symbol. *Educational Studies in Mathematics, 12*, 317-326.

Kieran, C. (1992). The learning and teaching of school algebra. In D. A. Grouws (Ed.), *Handbook of research on mathematics teaching and learning* (pp. 390-419). New York: National Council of Teachers of Mathematics & MacMillan.

Kieran, C. (2006). Research on the learning and teaching of algebra: A broadening of sources of meaning. In A. Gutiérrez & P. Boero (Eds.), *Handbook of research on the psychology of mathematics education: Past, present and future* (pp. 11-49). Rotterdam, The Netherlands: Sense Publishers.

Knuth, E. J., Alibali, M. W., McNeil, N. M., Weinberg, A., & Madison, A. C. S. (2005). *ZDM, 37*(1), 68-76.

Ministry of Education of Singapore. (2018). *Secondary mathematics syllabuses (draft)*. Singapore: Curriculum Planning and Development Division.

Rittle-Johnson, B., Matthews, P. G., & Taylor, R. S. (2011). Assessing knowledge of mathematical equivalence: A construct-modeling approach. *Journal of Educational Psychology, 103*(1), 85-104.

Warren, E., & Cooper, T. J. (2009). Developing mathematics understanding and abstraction: The case of equivalence in the elementary years. *Mathematics Education Research Journal, 21*, 76-95.

Yeo, K. K. J. (2019). Big ideas about equivalence in the primary mathematics classroom. In T. L. Toh & J. B. W. Yeo (Eds.), *Big ideas in mathematics* (Association of Mathematics Educators 2019 Yearbook, pp. 113-128). Singapore: World Scientific.

Chapter 11

Computational Thinking in Mathematics: To be or not to be, that is the question

Weng Kin HO, Chee Kit LOOI,

Wendy HUANG, Peter SEOW, Longkai WU

Computational Thinking is a paradigm for problem solving with the goal that problems and their solutions can be executed by a computer. Because of one's natural association of computer and computer programming, one is often misguided to think that computational thinking is solely reserved for the computer scientists and computer programmers. This chapter takes the stance that computational thinking is a generically useful way of thinking that is applicable across all disciplines, and in particular, mathematics. We highlight four design principles that mathematics teachers in Secondary Schools and Junior Colleges can apply to create lessons that promote computational thinking to forge mathematical ideas and enhance mathematics learning, which we term as "Math + C" lessons.

1 Introduction

During the 1960s, Seymour Papert and Alan Perlis painted the vision of how the world might look like with machine automation. While Perlis felt that programming should be integrated into higher education (Perlis, 1962), Papert envisioned that computational thinking, not programming, is the paradigm that should be introduced in K-12 education (Papert, 1980). Prior to his stint as co-director of the MIT Artificial Intelligence Laboratory from 1967 to 1981, Papert worked closely with Jean Piaget, the father of the theory of *constructivism* – learners construct new knowledge from their experiences of acquiring previous knowledge. Papert, inspired by Piaget, developed his learning theory called *constructionism* – learning can be enhanced when the learner is engaged in "constructing a meaningful product", which is an additional thesis augmenting Piaget's constructivism. In 1980, Papert published the book *Mindstorms: Children, Computers and Powerful Ideas.* In the introduction, he looked beyond the rise of the personal computers and the impact of home and work life to consider the question of "how computers may affect the way people think and learn". Papert coined this kind of thinking as *computational thinking* (Papert, 1980, p. 182) which he claimed would "enhance thinking and change patterns of access to knowledge".

Perhaps the world then was not yet ready for Papert's vision, and thus it was not to be taken up seriously. The term 'computational thinking' was to wait for almost three decades before Jeannette Wing, in her now-highly-cited article (Wing, 2006) published in *Viewpoint* a decade ago, brought up the term again – albeit with a different interpretation. Labelling computational thinking as a layman's "attitude and skill set", Wing advocated that solving problems inevitably invoke fundamental concepts of computer science: decomposition, recursion, abstraction, separation of concerns, etc. Thus, Wing's view about computational thinking is skewed towards computer science, in which she equates computational thinking with thinking like a computer scientist. Wing's interpretation of computational thinking – though overwhelmingly influential in the literature of computational thinking – seems to be far more restrictive than Papert's original idea.

1.1 *Definition of Computational Thinking adopted in this chapter*

In this section, we examine carefully Papert's original formulation of computational thinking while retaining the salient parts of Wing's interpretation. The authors hold the view that computational thinking is a problem solving paradigm which bears the unique characteristic of expressing problems and constructing solutions in such a way that a computer could execute. Here a computer refers to any agent (human or machine) which is able to carry out a finite set of instructions executed within the agent's physical capabilities. It is easy, at this point, to reject computational thinking as an essential way of thinking of everyone; after all, not everyone would become computer scientists or a computer programmer. We refute this by emphasising that "the goal is to use computational thinking to forge ideas" (Papert, 1996). Since computational thinking is intended to "change patterns of access to knowledge" (Papert, 1980), then it should be that everyone must be taught, in additional to many existing ways of thinking, to think computationally – the sooner the better.

As a paradigm for knowledge, computational thinking has its characteristics and dispositions – an important fact that Wing has already highlighted – which includes the following four fundamental features: Decomposition, Pattern recognition, Abstraction, Algorithmic design. Because these terms arise originally from the domain of computer science, it is natural for one to restrict computational thinking to computer science. This is where we return to Papert's theory of constructionism, which asserts that learner can improve his/her learning if there is learner's engagement in "constructing a meaningful product". Since every domain of knowledge speaks of a specific "meaningful product", we opine that computational thinking is domain-specific, and our sole concern in this chapter is *computational thinking in mathematics*, which we unpack as follows.

Decomposition is the process by which the mathematics problem is broken down into smaller sub-problems or sub-tasks. Commonly employed mathematics problem solving heuristics such as simplifying the problem, making suppositions, and trying out on smaller cases/numbers can be regarded as actions of problem decomposition. Through decomposition, the original problem, which at first was *complex*, now becomes manageable since each of the smaller sub-problems/tasks is tractable.

Pattern recognition is the action of looking out for common patterns, trends, characteristics or regularities in *data*. Mathematics is the study of regularity in structures, and thus lends itself well to this particular component. Indeed, patterns and the recognition of which occur whenever there are observable data involved in the problem. Data can come in many forms, e.g., numbers, vectors, shapes, mathematical structures, etc.

Abstraction is the process of formulating the general principles that generate these recognised patterns. In particular, abstraction takes place when a real world context problem is re-casted into a well-defined mathematics problem. Expressing everyday content using the language of mathematics is called *mathematising*.

Algorithmic design is the development of a precise step-by-step recipe or instructions for solving the problem at hand as well as problem similar to it. It is intended that such instructions can be executed in an 'insightless' manner by a human being or a machine. The theory of computability, a branch of mathematics, bears upon us that the set of instructions be of finite character, and in such a case, the problem we are solving is said to be *effectively calculable* or *computable.*

Before we leave this section, it is important for the reader to realize that there is no unique, universal and clear definition for the term 'computational thinking' so far in this literature of the domain of computational thinking. This problematic situation arises partly because of the varied interpretations of the term over different domains and disciplines (e.g., science, mathematics, etc.), and of the connections of these disciplines with computer science, as well as those issues centred on the degree of involvement of the computer. Some authors have also attempted to define computational thinking in terms of a taxonomy of practices focusing on the applications of computational thinking in mathematics and science (Weintrop et al., 2016). Thus, we do not claim that we have defined computational thinking in a universally accepted way; instead our focus here is to provide concrete guidance for teachers who wish to realize computational thinking in their mathematics classroom.

1.2 *Maths + C lessons*

For the mathematics teacher, computational thinking opens up a new portal for accessing knowledge. It is all too easy to say that infusing 'C' (Computational Thinking) into Maths (Mathematics) would do the trick. The fact is that bridging the gap between theory (constructionism, and in particular, computational thinking) and practice (actual classroom teaching) requires deliberate thought and careful planning. What would be useful for mathematics teachers would be a concrete set of lesson design principles that can help them in the process of planning for a meaningful lesson that promote constructionism in mathematics learning. We shall term such a lesson a "Math + C" lesson.

We want to emphasise that many researchers have already explored the use of computational thinking activities in teaching and learning mathematics; for this, consult the detailed literature review of such scholarly works in Barcelos et al (2018). There are also research works that look into the relationship between computational thinking and mathematics learning in the context of Singapore mathematics curriculum, e.g., link with coding, computational thinking and mathematics problem solving (Ho and Ang, 2015), and the ability of mathematics students thinking computationally (Ho et al, 2018). Though not a pioneer in the connecting computational thinking with mathematics education, our present approach is novel in that we develop "Maths + C" lesson design principles to help mathematics teachers create lessons that makes use of computational thinking to forge mathematical ideas – the subject of discussion in the next section.

2 Lesson design principles

Before we lay down the design principles, we surface certain practical issues a Math + C teacher has to deal with when writing a lesson plan. An awareness of these issues put the teacher in a better position in applying the four principles we are about to explicate.

2.1 *Practical issues to consider*

Firstly, it has to do with the selection of the topic, together with determination of those components of computational thinking one can make use of to teach the selected topic in a way that will be more impactful than the existing pedagogies. In other words, the question to ask is what role computational thinking plays in the lesson enactment. Does computational thinking serves to enhance learning and meets the teacher's *goals* for the lesson. Secondly, the teacher must consider the *available resources* to execute the lesson plan; for instance, does the lesson require specific computer software or programming language, and is there sufficient technological infrastructure to support computer-based learning? Thirdly, the teacher must be cognizant of *students' background knowledge* in relation to computational thinking/the use of computers, e.g., what levels of familiarity do students have with the terminologies of computational thinking, does the lesson require additional instructional time to equip students with computer-related/computational thinking skills? Fourthly, and perhaps most importantly, *teachers' belief, competency and readiness* in carrying out a Math + C lesson are also crucial points to consider.

2.2 *Lesson design principles*

The lesson design principles for infusing computational thinking into mathematics lessons anchor themselves on four key components of computational thinking, namely, decomposition, pattern recognition, abstraction and algorithm design.

   Complexity Principle: Does the topic/subtopic/concept give rise to sufficiently complex problem/situation? The problem/situation that is relevant to the learning of the topic/subtopic/concept must be complex enough so that decomposition of this main problem into sub-problems is a meaningful thing to do. If the problem or task is routine, e.g., it can be solved readily by a simple and well-known method, then decomposition of the problem will be unnecessary or superficial.

Data Principle: Does the topic/subtopic/concept manifest in many instances so that common traits/trends/patterns can be observed, quantified, stored and treated as data? The topic/subtopic/concept involves observable and quantifiable data that can be stored, treated, transformed and used.

Mathematics Principle: Does the topic/subtopic/concept give rise to a problem/situation that can be mathematised? By mathematising a problem/situation, we mean formulating the problem/situation in an abstract and precise manner using mathematics. Here mathematics is not restricted to just numbers, algebra, geometry, and takes on a more inclusive meaning to encompass any abstract concepts/structures that can be defined, represented, and reasoned about rigorously within some logical framework.

Computability Principle: Is there an effectively calculable solution to the mathematised problem/situation? Here, the word 'effective calculable' has a specific meaning as used in Computability Theory, i.e., there exists a solution to the problem/situation that can be implementable on a computer (be it a machine or human being) through a finite procedure (i.e., a finite sequence of instructions).

Based on the Papert's maxim that learning can be enhanced when the learner is engaged in constructing meaningful product, we propose that each Math + C lesson should get students engaged in *building some final product* that provides physical evidence of their understanding of the mathematics topic/concept/technique/result.

In order to keep the computer-science/technology overheads low, we shall only use the Microsoft EXCEL spreadsheet for simple coding. No other computer science pre-requisites are assumed for this chapter.

3 Sample Math + C lessons

In this section, we shall apply the design principles mentioned in the previous section to create Math + C lessons for Secondary Schools ('O' level Mathematics) and Junior Colleges ('A' level H1/H2 Mathematics) within the Singapore context.

We claim that the preceding design principles help mathematics teachers make important decisions in order to realize the goal of making a difference in teaching and learning mathematics *via* computational thinking. Because of limited space in writing, we shall give full details of the teacher's design trajectory for one example of number patterns in Section 3.1.1.

3.1 *Secondary Schools*

3.1.1 Number patterns

The first thing to realize is that the design principles need not be invoked in any order. Design principles phrased as questions facilitate self-dialogue for the teacher. Our hypothetical scenario involves a Secondary One Mathematics teacher who wishes to try using Computational Thinking to teach the more basic and introductory topics in the Singapore Mathematics syllabus for Secondary One (i.e., students aged 13). One of these topics is an 'O' level Mathematics syllabus topic that links (whole) numbers with algebra via a sequence of number patterns. This topic can be seen as a crucial link between Primary School Mathematics (where students are accustomed to using concrete representations of numbers) and Secondary School Mathematics (where students will learn algebra which is abstract in nature).

A typical problem involves a sequence of whole numbers that possesses some regularity in the generation of the next term from the previous one, and such a sequence is usually represented arithmetically or figuratively. Can this topic of number patterns offer a convenient platform for us to design a Math + C lesson? To answer this question, the teacher then appeals to the Data Principle: *Does the topic of "Number Patterns" manifest in many instances so that common traits/trends/ patterns can be observed, quantified, stored and treated as data?* There is clearly an affirmative answer to this question. Number patterns are just physical representation of numbers, and regularity is ubiquitous in number patterns so that such regularity is easily observed and described.

An example of a number pattern represented arithmetically is given in Figure 1(a), and one which is represented figuratively is given in Figure 1(b).

n	T_n	n	nth pattern	T_n
1	$1 = 1^2$	1		1
2	$1 + 3 = 2^2$	2	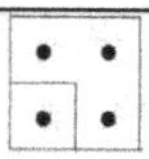	4
3	$1 + 3 + 5 = 3^2$	3	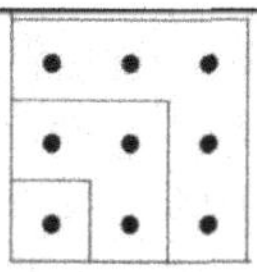	9
(a)		(b)		

Figure 1. Number patterns (a) arithmetically, (b) figuratively

In the 'O' level Mathematics Syllabus, it is expected that students know how to write down the closed formula for the nth term of a given sequence or number pattern. Although it is relatively easy for students to observe the recurrence relation that exists between the nth pattern and the $(n + 1)$th pattern, the job of writing down the closed formula for the general nth term, T_n, often proves challenging for most students. Many real-world contexts give rise to sequences which can be generated using very simple recurrence relation and yet no closed formulae exist for the general term – a fact that many students are not aware of. Additionally, for assessment purposes, most question items for this topic restrict the types of sequences $\{T_n\}_{n=1}^{\infty}$ to either a linear or a quadratic function of n. As a result, many students often have the misconception that any number sequence has to be one of these forms. Because of this misconception, it is quite common for students to employ specific forms (either $T_n = an + b$ or $an^2 + bn + c$) to solve the given number pattern problem, where the constants will be determined by a simultaneous system of equations for the unknown constants formed using the first few given terms of the sequence. This view taken by students oversimplifies the topic, undermines the importance of number sequences, and causes learning to be shallow.

Because of the aforementioned considerations triggered by the Data Principle, it is natural for the teacher to look for ways in which Computational Thinking can be considered as a means to help students in forging their understanding in the concept of number patterns.

The teacher's choice is further affirmed by the Computability Principle and the Mathematics Principle. Concerning the Computability Principle, the teacher asks himself or herself: "*Is there an effectively calculable solution to the mathematised problem of finding the nth term of the number pattern?*" Pertaining to this question, one recognises that recurrence relations can easily be coded in any programming language, and hence this problem of finding the nth terms of the sequence is certainly computable. This algorithmic approach always gives the exact (numerical) solution for the problem, even when no closed formulae exist – the prowess of numerical methods.

Concerning the Mathematics Principle, the teacher asks himself or herself: *Does the topic of Number Patterns give rise to a problem that can be mathematised?* To this question, we are assured that the computational thinking behind the use of the graphing features in EXCEL analysing the relation between n and T_n compels us to mathematise the given problem of establishing the mathematical relation between T_n and n.

Based on these two considerations triggered by the Computability Principle and the Mathematics Principle, the teacher would very likely think of a task for his or her students, whose specifications may be phrased as in Figure 2.

Task statement (Number patterns)

Employ the scatter plot and trend-line tools in EXCEL to explore the relationship between the index n and the general term, T_n, of the sequence, the first three terms of which are given in Figure 2(b). [Hint: Generate a table of values of T_n against n (for $n = 1,2,3,\ldots,10$) by identifying the relationship between T_{n-1} and T_n.]

Figure 2. Task statement (Number patterns)

Next, the students will be shown the 'final product' of the EXCEL worksheet that has been constructed to fill in the required table of values for the sequence (see Figure 3).

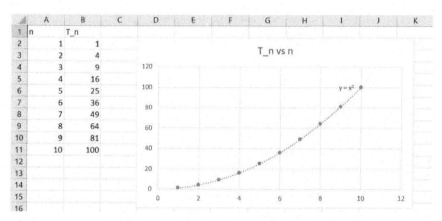

Figure 3. EXCEL worksheet depicting the relationship between n and T_n

A number of "just-in-time" EXCEL skills need to be taught in order to equip students with the ability of writing a mathematical formula that takes in values in specified cells as arguments, defining a sequence by recursion/recurrence relation, and obtaining the scatter plot and trend-line (linear/quadratic) for a bivariate table.

The teacher then proceeds to validate the lesson plan/task by invoking the design principles again.

Complexity Principle. This task is sufficiently complex in that (i) the sequence for the index n needs to be generated, (ii) the sequence T_n needs to be generated *recursively* by formulating the mathematical relationship between T_n and T_{n-1}, and (iii) the bivariate data (n, T_n) needs to be displayed in the form of a scatter plot together with the trend-line (polynomial fit). Each of these subtasks (i)-(iii) requires specific computational techniques and EXCEL skills to be taught before they can be accomplished by the student.

Data Principle. Students perceive the variables n and T_n as storage places for two different sets of data. The teacher can speak about the different data type or format an entry can take on within a given cell, such as general, currency, integer, string, etc. (Figure 3).

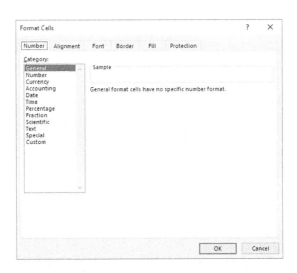

Figure 4. Different data type or format for cell data

In this case, both can be set to "General" or "Number". After the relationship between n and T_n is computed by EXCEL in the form of the equation of 'best-fitting' degree 2-polynomial, the students can also create a third column, say Column C, which computes the evaluation of the 'computed' polynomial on the input n stored in Column A. Column C contains data (yield from the closed formula) which will be verified to be equal to those calculated in Column B (yield from a recurrence relation). The teacher can then ask the students to describe the difference in the nature of the computational scheme employed in Columns B and C. Another question would be "Which scheme is the 'better' choice? Justify your choice."

Mathematics Principle. The recurrence relation between n and T_n can be precisely captured mathematically in an equation:

$$T_n = T_{n-1} + (2n - 1), n \geq 2.$$

For the computer to carry out the intended calculations, it is necessary that instructions given to the computer must be coded precisely; that is, every piece of syntax needed to carry out the instruction must be meaningfully and correctly coded. The 'meaning' part derives from the semantics of the required situation, and the 'correct' part comes from getting the accurate syntax coded as do-able instructions for the computer to execute.

Computability Principle. Importantly, the recurrence relation given in the above equation must be implementable in EXCEL codes. This can be done, say, by keying in the Cell B3 the syntax "`=B2+(2*A3-1)`", which effectively realizes the above mathematical equation. Dragging the formula in B3 vertically down for all the corresponding cells to the running 'indices' of Column A then automatically defines the corresponding formula which realizes the mathematical equation for each value of n in Column A; for instance, Cell B4 will be updated to "`=B3+(2*A4-1)`".

3.1.2 Partial fractions

Does a Math + C lesson always require a computer implementation? The answer is clearly 'No'. As pointed out, computational thinking is a way of problem solving process that includes the distinctive elements of decomposition, pattern recognition, abstraction and algorithmic design. Algorithm is often mistaken to be machine codes or computer programs; the truth is that an algorithm is just a finite set of precise instructions that can be carried out by a 'computer' in finite time. Human beings are equally capable of systematically treating and transforming data, and these abilities are often what one takes to solve a problem effectively – no machines involved!

In 'O' level Additional Mathematics, students are expected to perform partial fractions decomposition of a rational function $\frac{P(x)}{Q(x)}$, where $Q(x)$ is a polynomial of at most the third degree, and $P(x)$ another polynomial. For assessment purposes, there are certain restrictions to the following categories of the resultant partial fractions decomposition after performing long division to ensure that $\deg P(x) < \deg Q(x)$:

- Category 1: $\frac{A}{ax+b} + \frac{B}{cx+d}$, $ad - bc \neq 0$.
- Category 2: $\frac{A}{ax+b} + \frac{B}{(cx+d)} + \frac{C}{(cx+d)^2}$, $ad - bc \neq 0$.
- Category 3: $\frac{A}{ax+b} + \frac{Bx+C}{x^2+c^2}$, $c \neq 0$

A typical lesson suite for teaching this topic comprises episodes of various partial fractions decompositions that fall under the respective categories.

In this suggested lesson, we exploit the Algorithmic Design component to help students systematically perform the correct partial fractions decomposition for a given rational function. Crucially, we make use of flow charts. Figure 5 shows the task designed for this purpose.

Task statement (Partial fractions)

Design a flow chart that gives systematic instructions on how to perform partial fractions decomposition within the scope of the 'O' level Additional Mathematics syllabus, i.e., the given rational function is of the form $\frac{P(x)}{Q(x)}$, where

---

- $P(x)$ is a polynomial, and

- $Q(x)$ is a polynomial which takes *exactly one* of forms of factorization below:

  ○ $Q(x) = (ax + b)(cx + d)$; or

  ○ $Q(x) = (ax + b)(cx + d)^2$; or

  ○ $Q(x) = (ax + b)(x^2 + c^2)$.

Figure 5. Task statement (Partial fractions)

The final product of the required flow chart is shown in Figure 6, which we suggest to display at the end of this lesson. The skills of constructing flow charts must be taught before the students embark on the actual task.

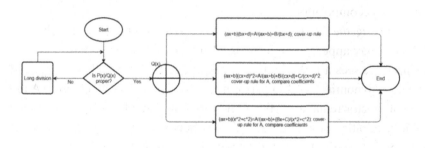

Figure 6. Sample flowchart for Partial Fractions Decomposition

We now validate the task using the four design principles.

Applying the Complexity Principle. It suffices to say that the task of carrying out the procedure of partial fractions decomposition correctly for a given rational function certainly involves many steps, and is far from being a straightforward task. To justify this claim, one cannot do so without simultaneously applying the Mathematical Principle. This example serves to illustrate the point that in certain situations there can be areas of overlap as far as the application of the four design principles.

The details of the validation using the Mathematical Principle is given later.

Applying the Data Principle. From the above detailed description, the salient data constitute of the various algebraic forms of the factorization of $Q(x)$, and information about the degrees of the polynomials $P(x)$ and $Q(x)$. From this example, we learn that not all data are numbers – data can also manifest in various forms and have different representations. If we wish to implement the subtasks using a computer, we may code a polynomial using a finite list of real numbers (its coefficients); e.g., the polynomial $-1 + x + 2x^2$ can be coded as $[-1,1,2]$. This coding technique allows us to make subsequent manipulation of the polynomial, e.g., addition/multiplication of polynomials. In our suggested lesson, there is no need to use this coding scheme. We merely make use of pseudocodes to give instructions (in natural language, say English) of what steps to perform during the partial fractions decomposition.

Applying the Mathematics Principle. The Mathematics involved in this task may appear to be just partial fractions decomposition. Looking deeper, one sees a full spectrum of *algebraic techniques* that are centred around polynomials: (AT1) perform long division of polynomials, (AT2) use of factor/remainder theorem for polynomials, (AT3) factorize cubic/quadratic polynomials, (AT4) make algebraic expansion of polynomials, (AT5) compare coefficients of polynomials when two polynomials are equal, (AT6) solve linear simultaneous equations, and (AT7) use 'cover-up rule'. The details are given below. The task can be decomposed systematically into the following steps:

Step 1. Check whether the rational function is proper, i.e., whether $\deg P(x) < \deg Q(x)$. If it is proper, then the rational function will be treated in Step 2. If not, one has to (AT1) perform long division/synthetic division of polynomials to obtain:

$$P(x) = Q(x)D(x) + R(x), 0 \le \deg R(x) < \deg Q(x).$$

Then the proper rational function so obtained will be $\frac{R(x)}{Q(x)}$.

For example, for the improper rational function $\frac{x^3+2x^2+7x+1}{(x+1)(x^2+4)}$, the corresponding long division would yield:

$$x^3 + 2x^2 + 7x + 1 = (x + 1)(x^2 + 4)1 + (x^2 + 3x).$$

Here $Q(x) = (x + 1)(x^2 + 4), R(x) = x^2 + 3x$ and $\deg(R(x)) = 2 < \deg(Q(x)) = 3$. The proper rational function obtained will be $\frac{x^2+3x}{(x+1)(x^2+4)}$.

Step 2. Perform partial fractions on the proper rational function so obtained from Step 1 by factorising $Q(x)$. This step requires the use of the Factor Theorem and the algebraic techniques needed for factorizing a cubic polynomial.

Step 3: Match the factorization obtained in Step 2 with *exactly one* the following forms:

Category 1: $Q(x) = (ax + b)(cx + d)$ (has no repeated factors); or

Category 2: $Q(x) = (ax + b)(cx + d)^2$ (has only one repeated factor of multiplicity 2); or

Category 3: $Q(x) = (ax + b)(x^2 + c^2)$.

Step 4: Match the selected form with the corresponding category for the correct form of the partial fractions decomposition:

Category 1: $\frac{A}{ax+b} + \frac{B}{cx+d}$, $ad - bc \neq 0$.

Category 2: $\frac{A}{ax+b} + \frac{B}{cx+d} + \frac{C}{(cx+d)^2}$, $ad - bc \neq 0$.

Category 3: $\dfrac{A}{ax+b} + \dfrac{Bx+C}{x^2+c^2}$, $c \neq 0$

The process of pattern matching based on Computational Thinking requires the learner to match the form of $Q(x)$ into one of the categories. At the juncture, it is important to realize that $ad - bc \neq 0$ ensures that for Category 1, $Q(x)$ has no repeated factor, and for Category 2, $Q(x)$ has only one repeated factor of multiplicity 2. The learner can be guided to appreciate the importance of this condition. For instance, the learner can be asked to construct two factors $ax + b$ and $cx + d$ for which $ad - bc = 0$. In this case, the student will realize that the respective requirements stipulated in Categories 1 and 2 are not met.

For example, if $Q(x) = (x + 1)(2x + 4)$ then $ad - bc = 1(4) - 2(2) = 0$ and in this case, $Q(x) = 2(x + 1)^2$ does not in fact have two distinct linear factors. If $Q(x) = (x + 1)(2x + 4)^2$ then again the condition $ad - bc \neq 0$ is violated and this is because $Q(x) = 4(x + 1)^3$ which would have a repeated factor of multiplicity 3, instead of 2.

Step 5: Employ the algebraic techniques of the matched category below to determine the constants A, B and/or C.

Category 1: Use 'cover-up' rule to calculate A and B. Alternatively, one can use the identity $P(x) \equiv A(cx + d) + B(ax + b)$ and making special choices of values of x, or comparing coefficients.

Category 2: Use 'cover-up' rule to calculate B and C. Additionally, one makes use of the identity $P(x) \equiv A(ax + b)(cx + d)^2 + B(cx + d)(ax + b) + C(ax + b)$ and comparing coefficients to find the value of A.

Category 3: Use 'cover-up' rule to calculate A. Additionally, one makes use of the identity $P(x) \equiv A(x^2 + c^2) + (Bx + C)(ax + b)$ and comparing coefficients to obtain the values of B and C.

For the running example, $P(x) = x^2 + 3x$ and so we have:

$$x^2 + 3x \equiv A(x + 1) + (Bx + C)(x^2 + 4),$$

where we can compare the coefficients of the quadratic and deduce that $A = -1, B = 2,$ and $C = 1.$

Applying the Computability Principle. In this lesson, no coding is involved. Indeed, we just make use of precise instructions written in the natural language for carrying out the different subtasks determined by the problem decomposition. Moreover, we organize these instructions using flow charts. The enthusiastic teacher may need to highlight the various features/symbols appearing in a flow chart, and their corresponding functions (see Table 1).

Symbol	Name	Function
	Start/End	An oval represents a start/end point of the program
	Arrows	A line connector that shows relationships between various shapes
	Input/Output	A parallelogram represents input or output
	Process	A rectangle represents a process
	Decision	A diamond indicates a decision
	Storage	A cylinder indicates a place for data storage

Table 1. Flow chart symbols and their functions

3.2 *Junior Colleges*

3.2.1 Maclaurin series

The topic of Maclaurin series is an opportunity for students to learn about the versatility of polynomials in the computation of infinitely differentiable functions, and the concept of approximation. Let $f(x)$ be an infinitely differentiable function defined on an open interval or the whole of the real line, and consider the Maclaurin polynomial $P_n(x)$ of the form $\sum_{k=0}^{n} \frac{f^{(k)}(0)}{k!} x^k$. The Maclaurin series is the limit of the polynomial as n tends to infinity, if it exists, which is equal to the function within the domain of convergence of the series. Usually, the bulk of the teaching and learning will be centred around finding the Maclaurin series of a given function by repeated differentiation (explicitly or implicitly) or using the Maclaurin series expansion of common functions (e.g., the binomial series for fractional index, the exponential function, the logarithmic function on restricted domains, the sine and cosine functions). Although the H2 Mathematics Syllabus explicitly states that the concept of "approximation" (with inverted commas) is related to this topic, it is unclear to the students and, sometimes the teachers, how "approximation" can be taught. Moreover, without using tertiary mathematics (e.g., radius of convergence, ratio test), it is difficult for a student appreciate the existence of the different domains of convergence for different functions.

The notion of "approximation" can be made meaningful when one calculates how close the approximating polynomials is to the given function $f(x)$, i.e., one must calculate the errors defined explicitly by the equation $\epsilon_n(x) := P_n(x) - f(x)$. Graphically, the errors can also be visualized by how near/far the curves of approximating polynomials to that of the given function. While the Graphing Calculators can display the graphs of the approximating polynomials and the given function, they rob the students of the opportunity to generate the values of $P_n(x)$, $f(x)$ and hence miss out on appreciating what the error function $\epsilon_n(x)$ is via the "table-of-values". The tabulation of the errors, say $\epsilon_3(x)$, can deepen students' understanding of approximation and estimation, an aspect that is often not tested in assessment items in examinations.

Computational Thinking creates the opportunity for students to handle values of functions, and error values as useful and handy data. The tedium of calculating these values can be taken over by the machine through the use of recursive definition (also known as recurrence) – a common feature in algorithm design and programming. Although recurrence relation is featured explicitly in H2 Mathematics Syllabus as the one of the ways of defining a sequence, the computational advantage of doing so over that of closed formulae is seldom mentioned to students. In this suggested lesson, the students are to employ the table of values over a user-declared closed interval $[a,b]$ for the values of the given function $f(x)$, the approximating polynomials $P_n(x)$ for n = 0, 1, 2, 3. The task statement is given in Figure 7.

Task statement (Maclaurin series)

Consider the function $f(x) = \sin(x)$ and the Maclaurin series expansion $\sum_{k=0}^{n} \frac{f^{(k)}(0)}{k!} x^k$ up to the various degrees of n = 0, 1, 2, 3. Using EXCEL worksheets, construct the a table of values to evaluate the given function f and each polynomial, over 101 values $a < x_1 < x_2 < \cdots < x_j < \cdots < b$ equally spaced in the closed interval $[a,b]$, where a and b are values entered by the user. Tabulate the error of estimation for using the Maclaurin series up to degree 3.

Figure 7. Task statement (Maclaurin series)

The table of values template is shown to the students (see Table 2).

j	x_j	$\sin(x_j)$	$P_0(x_j)$	$P_1(x_j)$	$P_2(x_j)$	$P_3(x_j)$	$\epsilon_3(x)$
0	a						
⋮	⋮						
101	b						

Table 2. Table of values template

Next, the students would be shown the 'final product' of the EXCEL worksheet that has been constructed to fill in the table of values that varied according to the values of a and b which are to be entered by the user (see Figure 8).

In Figure 8, the values of a and b selected by the user are 0 and 2 respectively. Because there are no x^2 terms in the 'quadratic' Maclaurin polynomial $P_2(x_j)$, this polynomial coincides with the linear polynomial so that there are only the graphs of the three polynomial approximations and the original function. The teacher then demonstrates a change of values for a and b to show that the EXCEL worksheet is dynamic.

We again apply the four design principles to construct the lesson plan, justifying the suitability of the activities/subtasks that have been chosen.

Applying the Complexity Principle. The task at hand is fairly complex: (1) the input span of x values, namely the x_j's, must first be generated given the values of a and b entered by the user, (2) the Maclaurin polynomial approximations is to be worked out correctly for the varying degrees $n = 0, 1, 2,$ and 3, (3) the errors $P_3(x) - f(x)$ are to be computed for the input range of values of x, and (4) the graphs of the three approximating polynomials and the original function need to be displayed. The teacher points the students towards decomposing the original task into smaller sub-tasks which they can handle one after another.

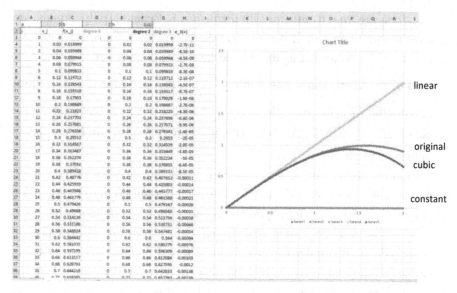

Figure 8. EXCEL worksheet showing the graphs of the original (sine) function and three Maclaurin polynomial approximations in [0,2]

Applying the Data Principle. Because of the 101 points evenly spaced in the closed interval [a,b], there is quite a lot of numerical data to be handled. This lesson intentionally infuses computational thinking into an otherwise mundane routine of repeated differentiation so that students have the opportunity of transforming data to yield meaningful new data, e.g., errors.

Applying the Mathematics Principle. The notion of "approximation" is now made very concrete as the error is not perceived as a function, i.e., it depends on the input value of x as well as the degree of the Maclaurin polynomial $P_n(x)$ which approximates $f(x)$. The teacher then points the students to the EXCEL worksheet they have constructed themselves to answer a few important mathematical questions: (1) As the degree n increases, what can you say about the accuracy of the approximation? (2) For a fixed value of n, what can you say about the accuracy of the approximation as x increases, i.e., as it moves away from 0? (3) Why are polynomials chosen as approximations, and not some other functions?

Applying the Computability Principle.   Because polynomials are computable functions, it is clear that the computability principle is valid in this case. Note that as the degree increases from $n - 1$ to n, the next term of the Maclaurin series, namely $\dfrac{f^{(n)}(0)}{n!}$, can be added to the previous polynomial $P_{n-1}(x)$ to yield the new value for the polynomial of the next higher degree, i.e., $P_n(x)$. This is where recursion can be employed in the calculations.

Other than our pre-occupation with whether a function is computable, we should add in two important remarks regarding computation which *strictly speaking do not concern computability*.

Firstly, one recognizes that "round-off" errors are always present whenever the computer tries to compute a real number. A computer in reality cannot compute a real number exactly in the sense that the decimal representation of certain real numbers does not terminate. Rounding-off errors can occur and 'snowball' in an actual computation. In our example here, the computation of the terms $\dfrac{f^{(n)}(0)}{n!} x^n$ already incur some rounding-off errors, and these will build up to an error in the final computation of the Maclaurin's series.   Students should be guided appreciate the presence of numerical errors arising from these rounding-off errors, and how improvement in the algorithms may help improve the accuracy of these calculations.

Secondly, the teacher can draw the learner's attention to issues concerning the efficiency of computation. In this example of calculating the Maclaurin's series, one needs to perform repeated evaluations of polynomials. In order to calculate the series more efficiently, one must use "nested algorithms" so as to reduce the number of arithmetic operations that have already been identified by the Complexity Principle, and the Mathematics Principle. In this example, the calculation of $P_2(x)$ makes use of $P_1(x)$ via the equation $P_2(x) = P_1(x) + \dfrac{1}{2}x^2$. Hence the equation in CELL(F3) of Figure 8 that evaluates $P_2(x)$ for the input x stored in CELL(B3) can be written as "=E3+0.5*(B3)^2", where CELL(E3) already holds the value of

$P_2(x)$. In this way, we see that Computational Thinking has an important impact in the teaching and learning of mathematics because it creates a difference in impacting the associated mathematics, e.g., in this case, our approach via computational thinking makes a difference in the student's understanding of Maclaurin's series as an approximation to the actual function.

3.2.2 Discrete Random Variables

In H1/H2 Mathematics, a key example of a discrete random variable is one that has a binomial distribution. While students are expected to know the formula for the probability mass function, i.e., $P(X = r) = \binom{n}{r} p^r (1-p)^{n-r}$ (where n is the number of independent trials, and p is the probability of success for each such trial), students often have difficulties understanding that the distribution function serves as only a model to describe the randomness of the event in question. Because the formula of the distribution function is deterministic in producing the probability of occurrence of the event '$X = r$' for any given integral value of r in $\{0, 1, 2, ..., n\}$, students often missed out on observing the randomness of each instantiation of X. This problematic situation is not helped by using the Graphing Calculator which merely calculates the probability of the binomial event '$X = r$' with the input parameters n and p.

In order to help students appreciate the randomness of the binomial event and understand the distinction between the theoretical relative frequencies (i.e., the probabilities given by the distribution) and the empirical relative frequencies (i.e., taking frequency of occurrences over the total frequency), one adopts a 'frequentist' approach. We implement this approach by asking students to perform the experiments and observe the random outcomes of the event in question. Although such experiments can be done 'by hand' individually, such a classroom implementation is often time-consuming, tedious and hard for the teacher to manage. Moreover, the number of experiments performed by each student is often small and hence the theoretical model may not 'fit' well with empirical frequencies obtained from these experiments, thus defeating the purpose of conducting such experiments. We propose that computational thinking when infused into the mathematics lesson can enhance teaching and learning of this topic. In this suggested lesson plan, the students' task is to design EXCEL simulations of a binomial event for a large number of times. The task statement is printed in Figure 9.

Task statement (Discrete random variables)

Simulate 5 sets of 1000 runs of a binomial experiment in which the probability distribution function of random variable X is given by $X \sim B(10, \frac{1}{2})$. Produce 5 corresponding histograms that display the (relative) frequencies of occurrence for each instance $X = r$ for $r = 0, 1, 2, \ldots, 10$. Compare and contrast the shape of the histograms obtained with that given by the binomial distribution function.

Figure 9. Task statement (Discrete random variables)

At the beginning of the lesson, the 'final product' of the EXCEL worksheet would be displayed on the screen (see Figure 10), and a life demonstration will be conducted by pressing the F9 button several times to generate a few sets of 1000 simulation of the binomial event. The histogram, which is projected on the screen, is updated for each simulation. The teacher will briefly explain the features of the EXCEL worksheet as well as the task requirement. The students' task is then to replicate the EXCEL worksheet so that it possesses all the required specifications of the displayed 'final product'. The final product is left on the screen for students' reference.

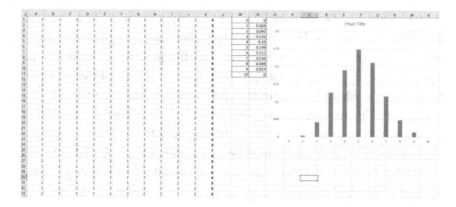

Figure 10. A single EXCEL simulation of the binomial event.

We again apply the four design principles to construct the lesson plan, justifying the suitability of the activities/subtasks that have been chosen.

Applying the Complexity Principle. The problem as expressed in Figure 9 is of sufficient complexity because it is not immediately clear how a simulation of a single Bernoulli trial with $p = \frac{1}{2}$ can be performed in EXCEL, let alone 1000 runs of 10 independent trials. The Complexity Principle bears upon the teacher to create suitable scaffolding to help students accomplish the intended task. There are two things to take care of when guiding students to practise decomposition. Firstly, the students must recognize what are the sub-tasks/sub-problems that are useful in solving the final problem and are within reach of their current abilities. Secondly, the teacher must provide just-in-time skills/knowledge to perform some of these sub-tasks. In this case, it is important to teach students the EXCEL command `rand()` which generates a random number in the interval [0,1], and perhaps also to introduce the ceiling function `ceiling( , )`. A possible sub-task that makes use of this specific knowledge would be for students to simulate a Bernoulli event with the two outcomes denoted by '1' and '2', each with probability of occurrence $p = \frac{1}{2}$. This simple sub-task can be achieved by the syntax `ceiling(2*rand(),1)`.

Applying the Data Principle. The seemingly trivial fact that the Bernoulli trial results in two outcomes that can be coded using the numbers '1' and '2' already suggests that the problem at hand has its manifestation in several scenarios where numerical data (different kinds of information about the binomial experiment) are heavily involved. To begin with, in 10 independent repetitions of the Bernoulli event the student would need to find out the total number of 'successes' observed, e.g., the number of '1' that turned up, per simulation of 10 trials. In the H1/H2 Mathematics syllabus, this random variable, which is denoted by X, has a theoretical distribution given by the binomial distribution $B(10, \frac{1}{2})$. To appreciate that this distribution is indeed a model that describes the phenomenon of randomness in this kind of event, the student has to perform a number of computations on a set of data obtained from one simulation to determine the empirical probability (i.e., the relative frequency of 'successes'), and to compare this with the theoretical one. Now, the teacher needs to apply the Complexity Principle again to design suitable subtask(s) for the students in order to achieve this desired learning outcome. In particular, the teacher needs to teach the concept of arrays in EXCEL and the array sum command sum(:) before assigning the subtask(s).

Applying the Mathematics Principle. Having (part of) the problem/situation quantified using numbers alone does not constitute mathematising. Indeed, the key action in mathematising is to draw out the mathematical essence of the problem/situation, which in turn allows the learner not only to model this particular situation but also to generalise to other situations. Mathematising involves both setting up of the concrete variables/parameters involved in the simulation at hand as well as getting the learner to lay down the assumptions needed during the process of solving the problem. In this lesson, the teacher can then teach students about the concepts of empirical frequency and theoretical frequency, and illustrate how these frequencies may be represented, e.g., . using a histogram. Revisiting both the Complexity and Data Principles, the teacher would plan to teach his/her students to create a histogram associated to a given table of (empirical/theoretical) frequencies using

EXCEL. Mathematising the current situation allows the students to transfer the same set of knowledge/skills to another situation, e.g., to investigate other probability distributions, such as the Poisson distribution, or to study other concepts/results, such as the Central Limit Theorem.

Applying the Computability Principle. Note that all the calculations described above are implementable by a computer, either using a software such as EXCEL or even a human being. Perhaps, an interesting point to note is that the random number generator is computable, and hence completely deterministic – nothing is random about it. The output appears to be 'random' partly because of time-based inputs and the ergodic nature of the algorithm.

4 Conclusion

"To be or not to be, that's the question" are words taken from the play *Hamlet* by William Shakespeare. These are the words that begin a famous speech by Prince Hamlet in which he contemplates whether to commit suicide as an escape from his troubles. To some extent, though of lesser gravity, there is an analogous contemplation in our context. Mathematics teachers who see some value in computational thinking are often caught in a dilemma, i.e., they ask, "Though I appreciate the potential benefits and powers of computational thinking, how can I make use of it in my lesson to help my students forge their ideas and concepts in mathematics? Should I do it or not?" Our preceding discussion answers this contemplative question in the affirmative. Based on Papert's theory of constructionism and the four characterising features of computational thinking, we propose lesson design principles that guide mathematics teachers in choosing the suitable topics, designing and validating their lesson plans with the aim that the learner constructs mathematically meaningful products. In our future work, we propose to study the versatility and effectiveness of these design principles in actual classroom implementations.

Acknowledgement

Our research work reported here is supported by the OER10/18 LCK (How to Bring Computational Thinking (CT) Into Mathematics Classrooms: Designing for Disciplinary-specific CT).

References

Barcelos, T., Munoz, R., Villarroel, R., Merino, E., and Silveira, I. (2018). Mathematics Learning through Computational Thinking Activities: A Systematic Literature Review. *Journal of Universal Computer Science, 24*(7), 815-845.

Ho, W. K., and Ang, K. C. (2015). Developing Computational Thinking Through Coding. In Yang W.-C., Meade D., and Liu, C. (Eds) *Proceedings of 20th Asian Technology Conference in Mathematics*, pp. 73-87. Mathematics and Technology, LCC.

Ho, W. K., Huang, W., and Looi, C. K. (2018). Can Secondary School Mathematics Students Be Taught Computational Thinking? In Yang W.-C., and Meade, D. (Eds) *Proceedings of 23rd Asian Technology Conference in Mathematics*, pp. 63-77.

Papert, S. (1996). An Exploration in the Space of Mathematics Educations. *International Journal of Computers for Mathematical Learning, 1*(1), 95-123.

Papert, S. (1980). *Mindstorms: children, computers, and powerful ideas.* Basic Books, Inc. New York, NY. ISBN:0-465-04627-4.

Perlis, A. (1962). The computer in the university. In Greenberger, M., (Ed) *Computers and the world of the future.* MIT Press, Cambridge, pp. 180-219.

Weintrop, D., Beheshti, E., Horn, M., Orton, K., Jona, K., Trouille, L., and Wilensky, U. (2016). Defining Computational Thinking for Mathematics and Science Classrooms. *J. Sci. Educ. Technol., 25*:127–147.

Wing, J. (2006). Computational Thinking. *Communications of the ACM, 49*(3), 33-35.

Contributing Authors

Lu Pien CHENG is a Lecturer in the Mathematics and Mathematics Education Academic Group at the National Institute of Education, Nanyang Technological University, Singapore. She received her PhD in Mathematics Education from the University of Georgia (U.S.) in 2006. She specializes in mathematics education courses for primary school teachers. Her research interests include the professional development of primary school mathematics teachers, task design in mathematics education and cultivating children's mathematical thinking in the mathematics classrooms.

Ban Heng CHOY, a recipient of the National Institute of Education (NIE) Overseas Graduate Scholarship in 2011, is currently an Assistant Professor in the Mathematics and Mathematics Education Academic Group in NIE. Prior to joining NIE, he taught secondary school students mathematics for more than 10 years, and was the Head of Department (special Projects) before he joined the Curriculum Planning and Development Division as a Curriculum Policy Officer. He holds a Doctor of Philosophy (Mathematics Education) from the University of Auckland, New Zealand, specializing in mathematics teacher noticing. Ban Heng was awarded the Early Career Award at the 2013 Mathematics Education Research Group of Australasia Conference in Melbourne for his excellence in writing and for presenting a piece of mathematics education research. Currently, Ban Heng is the Principal Investigator of a research which looks into the professional development of primary mathematics teachers; the research is funded by the MOE Academy Funding Grant.

Weng Kin HO is an Associate Professor in the Mathematics and Mathematics Education Academic Group at the National Institute of Education, Nanyang Technological University, Singapore. He received his Ph.D. in Computer Science from The University of Birmingham (UK) in 2006. His research interests also include domain theory, exact real arithmetic, category theory, algebra, real analysis and applications of topology in computation theory. He looks into the use of technology in teaching and learning mathematics, mathematics through computational thinking approaches, use of video technology in flipped classroom pedagogies, problem solving in mathematics and the use of history of mathematics.

Wendy HUANG is a research associate at the National Institute of Education, Nanyang Technological University, in Singapore. Her current work centers on computer science education and computational thinking, alongside a passion for building equitable and inclusive learning environments for historically non-dominant students in computing. Previously, she was the program manager for the Scheller Teacher Education Program (Massachusetts Institute of Technology, USA), a computer science and mathematics teacher (Teach for America), a teacher educator, and a developer of CS and STEM educational software and curricula (e.g., StarLogo Nova, Building Math). She has a M.S. degree in Teaching from Pace University (New York, USA) and a B.A. degree in Computer Science from Wheaton College (Illinois, USA).

Berinderjeet KAUR is a Professor of Mathematics Education at the National Institute of Education, Nanyang Technological University, Singapore. She holds a PhD in Mathematics Education from Monash University in Australia. She has been with the Institute for the last 30 years and best described as the doyenne of Mathematics Education in Singapore. In 2010, she became the first full professor of Mathematics Education in Singapore. She has been involved in numerous international studies of Mathematics Education and was the Mathematics Consultant to TIMSS 2011. She was also a core member of the MEG (Mathematics Expert Group) for PISA 2015. She is passionate about the development of mathematics teachers and in turn the learning of mathematics by children in schools. Her accolades at the national level include the public administration medal in 2006 by the President of Singapore, the long public service with distinction medal in 2016 by the President of Singapore and in 2015, in celebration of 50 years of Singapore's nation building, recognition as an outstanding educator by the Sikh Community in Singapore for contributions towards nation building.

Barry KISSANE is an Emeritus Associate Professor in Mathematics Education at Murdoch University in Perth, Western Australia. For almost thirty years, he taught primary and secondary mathematics teacher education students at Murdoch University, except for a period for which he was Dean of the School of Education and an earlier period working and studying at the University of Chicago. His research interests in mathematics education include the use of technology for teaching and learning mathematics and statistics, numeracy, curriculum development, popular mathematics and teacher education. He was written several books and many papers related to the use of calculators in school mathematics, and published papers on other topics, including the use of the Internet and the development of numeracy. Barry has served terms as President of the Mathematical Association of Western Australia (MAWA) and as President of the Australian Association of Mathematics Teachers (AAMT). He has been a member of editorial panels of various Australian journals for mathematics teachers for around 40 years, including several years as Editor of The Australian Mathematics Teacher. A regular contributor to conferences for mathematics teachers throughout Australasia and elsewhere, he is an Honorary Life member of both the AAMT and the MAWA.

Soo Jin LEE is an associate professor of mathematics education at Korea National University of Education, South Korea. After she completed her Ph.D. in mathematics education at University of Georgia in 2010, she had served as an assistant professor of mathematics education at department of mathematical sciences at Montclair State University in United States for three years. She conducts research on how teachers and students reason about quantities in the upper elementary and middle grades mathematics. In particular, she has been studying how teachers' and students' flexibilities with multi-level unit structures and various partitioning strategies could support their advanced mathematical reasoning, which includes fraction arithmetic and proportional relationship. She is currently serving as an editorial board member of the Mathematical Thinking and Learning, and a reviewer of peer-reviewed international journals: Journal of Mathematical Behavior, Journal for Research in Mathematics Education, and International Journal of Science and Mathematics Education.

Kam Ming LIM holds the appointment of the Registrar of the National Institute of Education and Deputy Divisional Director of the Office of Academic Administration and Services at the National Institute of Education, Nanyang Technological University, Singapore. He is also an Associate Professor with the Psychological and Child & Human Development Academic Group. He was conferred the Public Administration Medal (Bronze) by the President of the Republic of Singapore in 2015. He is currently the President of the Educational Research Association of Singapore, Board member of the Asia-Pacific Educational Research Association, and Council member of the World Education Research Association (2015 – 2019).

Lee Hean LIM is an Associate Professor at the Policy, Curriculum and Leadership Studies Academic Group at the National Institute of Education, Nanyang Technological University, Singapore. She held positions as school vice-principal and head of mathematics department prior to receiving an NTU scholarship to pursue her interest in the theory and practice of mentoring, leadership and management. She has been involved in postgraduate and in-service curriculum conceptualization, design and delivery of courses for professional development.

Chee-Kit LOOI obtained his PhD in Artificial Intelligence from the University of Edinburgh. His doctoral thesis is in the area of teaching novices programming. He was the founding Head of the Learning Sciences Lab, the first research centre devoted to the study of the sciences of learning in the Asia-Pacific region. His research interests are in the areas of intelligent educational systems, computational thinking, and seamless learning. His research in the area of mathematics education includes the design and development of digital mathematics manipulatives which have been made available to all secondary schools in Singapore, the design of an intelligent tutoring system for word problem-solving, and research using SimCalc for teaching rate of change to middle school students.

Leng LOW has taught mathematics in secondary schools for 27 years and is currently a Master Teacher at the Academy of Singapore Teachers. Her interest in catering to the different needs of the students saw her partnering teachers in networked learning communities and in professional development courses. As a Master Teacher, she strives to engage in continual professional growth through meaningful collaboration with teachers so as to deepen students' conceptual understanding of mathematics and to nurture in them the joy of learning. She has presented in both local and international conferences in the learning of mathematics and mentoring.

Peter SEOW is a Research Scientist in the Office of Education Research at the National Institute of Education, Nanyang Technological University. He works on developing pedagogies for teaching computing and development of Computational Thinking skills for students in schools. He is interested in how the use of Physical Computing devices can develop computational modelling skills and computational thinking competencies to solve problems. His other research areas includes the development of partnerships among schools, the Science Centre and student welfare service agencies to design, test, and implement new learning experiences particularly for lower-track students in schools.

Jaehong SHIN is an associate professor of mathematics education at Korea National University of Education. He obtained his PhD from University of Georgia, U.S. in 2010. The theme of his PhD dissertation was to investigate middle school students' fraction knowledge as a foundation for their construction of algebraic knowledge. His research interests now include all K-12 students' construction of algebraic knowledge, with a particular focus on proportional reasoning, and covariational reasoning. He is currently serving as an editor in chief of one the two most cited mathematics education research journals in Korea the School Mathematics, and a reviewer of peer-reviewed mathematics education journals including International Journal of Science and Mathematics Education.

Tin Lam TOH is an Associate Professor and currently the Deputy Head of the Mathematics and Mathematics Education Academic Group in the National Institute of Education, Nanyang Technological University, Singapore. He obtained his PhD from the National University of Singapore in 2001. He publishes papers in international scientific journals in both Mathematics and Mathematics Education, and is doing research in both areas. He has been involved in several mathematics education research projects on mathematical problem solving and comics for mathematics instruction.

Cherng Luen TONG is a research assistant at the National Institute of Education, Nanyang Technological University, Singapore. He has a Bachelor of Engineering from the Nanyang Technological University of Singapore. Prior to joining the Institute, he was a secondary school mathematics teacher for more than ten years. He is interested in the use of technology in capturing classroom activities and also quantitative analysis.

Diem H VUONG was one of the three recipients of the 2018-2019 Fulbright Distinguished Awards in Teaching fellowship from the United States. Her experience in teaching Math extends from the classroom to coaching teachers, to working at the community college level. She is well known for her effectiveness with students in gaining content mastery in mathematics. She had encouraged student success by utilizing a variety of teaching techniques to include employing the Biological Sciences Curriculum Study (BSCS) 5E Instructional Model, to providing a variety of alternatives to measure student achievement.

Lai Fong WONG has been a mathematics teacher for over 20 years. For her exemplary teaching and conduct, she was given the President's Award for Teachers in 2009. As a Head of Department (Mathematics) from 2001 to 2009, a Senior Teacher and then a Lead Teacher for Mathematics, she set the tone for teaching the subject in her school. Recipient of a Post-graduate Scholarship from the Singapore Ministry of Education, she pursued a Master of Education in Mathematics at the National Institute of Education. Presently, she is involved in several Networked Learning Communities looking at ways to infuse mathematical reasoning, metacognitive strategies, and real-life context in the teaching of mathematics. Lai Fong is active in the professional development of mathematics teachers and in recognition of her significant contribution toward the professional development of Singapore teachers, she was conferred the Associate of Academy of Singapore Teachers in 2015 and 2016. She is currently seconded as a Teaching Fellow in the National Institute of Education, Nanyang Technological University, Singapore, and is also an executive committee member of the Association of Mathematics Educators.

Longkai WU is a Research Scientist in the Office of Education Research at the National Institute of Education, Nanyang Technological University. He works on bringing Computational Thinking (CT) into the Mathematics Classroom by designing pedagogies for teaching computing and developing of Computational Thinking skills for students in schools. His other research areas includes diffusion of teaching innovations, Information and Communication Technologies and Leadership, Learning Technologies, Scaling and Translation.

Von Bing YAP obtained a B. Sc. (Hons) in Mathematics and a M. Sc. in Applied Mathematics, both from the National University of Singapore (NUS), and then a Ph. D. in Statistics from the University of California. Since 2004, he has been teaching at the Department of Statistics and Applied Probability in NUS, where he is an Associate Professor. His main interest is applied statistics, the interface between mathematical theory and science. Other fields that fascinate him include probability, molecular evolution, comparative genomics, ecology, and the philosophy and practice of teaching abstract concepts.

Joseph B. W. YEO is a lecturer in the Mathematics and Mathematics Education Academic Group at the National Institute of Education, Nanyang Technological University, Singapore. He is the first author of the *New Syllabus Mathematics* textbooks used in many secondary schools in Singapore. His research interests are on innovative pedagogies that engage the minds and hearts of mathematics learners. These include the use of an inquiry approach to learning mathematics (e.g. guided-discovery learning and investigation), ICT, and motivation strategies to arouse students' interest in mathematics (e.g. catchy maths songs, amusing maths videos, witty comics, intriguing puzzles and games, and real-life examples and applications). He is also the Chairman of Singapore and Asian Schools Math Olympiad (SASMO) Advisory Council, and the creator of Cheryl's birthday puzzle that went viral in 2015.

Joseph Kai Kow YEO is a Senior Lecturer in the Mathematics and Mathematics Education Academic Group at the National Institute of Education, Nanyang Technological University, Singapore. Before joining the National Institute of Education in 2000, he held the post of Vice Principal and Head of Mathematics Department in secondary schools. As a mathematics educator, he teaches pre- and in-service as well as postgraduate courses in mathematics education and supervises postgraduate students pursuing Masters degrees. His publication and research interests include mathematical problem solving at the primary and secondary levels, mathematics pedagogical content knowledge of teachers, mathematics teaching in primary schools and mathematics anxiety.